On Fuzzy Cluster Discovery

Towards Unsupervised Fuzzy Discovery of an Undetermined Number of Clusters for a Generic Non-Relational Notion of Homogeneity

ARASH ABADPOUR

Arash Abadpour (arash@abadpour.com) is with Fio Corporation.

The author acknowledges that this text has been submitted for publication and that this file will be deleted upon publication.

Contents

List of Figures

Abstract

Fuzzy clustering is a mature field of study which involves the unsupervised grouping of a number of observations into homogenous clusters. This prerequisite for a diverse set of problems in many fields of computer and other sciences has been traditionally concerned with notions of homogeneity which are relational. Additionally, fuzzy clustering took off based on the assumption that one is going to know, with a great deal of certainty, how many clusters are present in an input set of data items, what is denoted as the C number in this work. As clustering algorithms made significant progress towards separating a known number of clusters from the data, it was observed that this assumption is in fact at the same time both non-trivial and critically important. Subsequent works, including the important category of VAT algorithms, attempted, with various degrees of success in many cases, to produce a usable estimate for the number of clusters prior to any clustering having happened. Nevertheless, the determination of the number of clusters appropriate for an arbitrary problem instance is a challenge to date, specifically when a non-relational and generic notion of homogeneity is applicable. In this paper, we argue that the number of clusters present in a set of data items, that corresponds to a physical phenomenon, is in fact not a deterministic number to be known. In other words, the problem is *not* to find "the C clusters" present in a given set of data items, but, instead, to discover clusters in the data and to allow the user to select how many of the discovered clusters are applicable to their case. In other words, C is not an entity to be sought for, but, we argue that, it is the mirage of certainty which has afflicted the fuzzy clustering literature. Thus, we eliminate C from the model and utilize the results of multiple independent executions of a robustified single-cluster clustering model which we aggregate using a non-relational class-independent framework. This process results in many clusters, for each one of which associated prominence and weight indicators are calculated.

Chapter 1

Introduction

People tweet, real estate agents describe properties, and cameras capture image and video signals. These are only a few examples for the plethora of processes which generate huge volumes of *data items* in the modern world. These oceans of data items need to be classified in order to be of practical value, and this important task is carried out based on the *notion of homogeneity* which is inherent in the physical phenomenon to which the data items correspond. Half a billion tweets are published every day [1], 93,000 residences were sold in the Greater Toronto Area in 2014 [2], and it is estimated that 1 trillion digital photos were taken in 2015 worldwide [3]. Nevertheless, one can refer to "types of tweets", "classes of dwellings" and "categories of photographs". These informal references to the concept of unsupervised clustering are only meaningful because of the relevance of homogeneity.

Homogeneity, albeit appearing to be a trivial concept, is a misleading mathematical entity. In fact, one may succumb to the temptation of making the convenient assumption that homogeneity is an affair between a pair of data items. Simple and intuitive phenomena in fact act as *invalid* assurance that a relational framework is the *one and only* framework in which data items can interact. A more careful inspection, however, provides evidence for the contrary. And this evidence in fact belongs to *problem class*es which are as close to home as Euclidean and other relational problem classes are. For example, one can consider the problem class which is concerned with clustering data items which belong to $\mathbb{R}^2$ to lines on this plane. In this context, it is meaningless to refer to the *similarity* of any pair of data items. In other words, the distance between $x_1, x_2 \in \mathbb{R}^2$ is absolutely irrelevant to the likelihood that x_1 and x_2 belong to the same linear cluster.

Figure 1.1 provides a symbolic representation of the inadequacy of the relational model for a

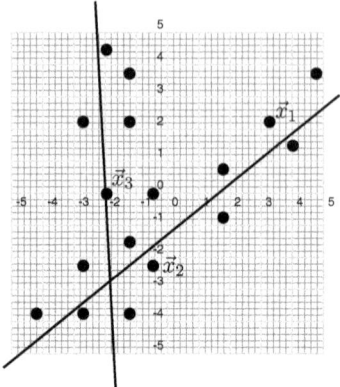

Figure 1.1: Symbolic representation of the inadequacy of the relational model for a generic notion of homogeneity.

generic notion of homogeneity, here for the problem class concerned with finding lines on a 2D plane. In this particular *problem instance*, two clusters exist in the input set of data items, as denoted by the two lines. Whereas $\vec{x}_2$ is "closer", in Euclidean terms, to $\vec{x}_3$ than to $\vec{x}_1$, it is in fact more likely to belong to the same cluster as $\vec{x}_1$ than $\vec{x}_3$. In other words, the Euclidean relationship between the data items is not a valid predictor of their co-clusterability.

Hence, clustering a set of data items into homogenous sets is governed by the distances between these data items and the clusters, which are yet to be discovered, and not the distances between the data items. The latter is in effect an *unimportant* piece of information which in cases may be *meaningless*. This approach is a solid departure from the limited domain of relational clustering and an ascension to a broad perspective to the data clustering problem. As will be discussed in this paper, we couple this step with a class-independent framework which allows for an economical approach to data clustering, in which one can afford to reuse not only the mathematics but also the actual computer code which discovers cluster representations in a set of data items of an arbitrary model.

It is reasonable to assume that the primary aim of a data clustering algorithm is to discover homogenous sets in a given set of data items. This problem statement, however, downplays the significance of one aspect of the solution to the data clustering problem, i.e. of the number of those clusters. In effect, it is one problem to be able to find C clusters in a set of data items, and it is an entirely different problem to be able to estimate C. The interdependency between "the clusters" and "the number of clusters" is evident in the history of fuzzy clustering. Fuzzy C-Means (FCM) is arguably the groundbreaking algorithm in this field, and, nevertheless, it makes the assumption that C is known. Visual Assessment of Cluster Tendency (VAT) is another important pillar of this field, and, in fact, although VAT was developed many decades after FCM, it is essentially a clever attempt to provide the key piece of information that FCM has been lacking, i.e. C.

The literature suggests that without having an estimate for C, the output of FCM, and a majority of other algorithms in the field, are suboptimal and that robust estimation of C is still beyond reach. In this work, we propose an alternative framework which practically *eliminates* C from the equation. In other words, we question the necessity for the existence of C in the fuzzy clustering framework and propose a model which is independent of it.

The rest of this paper is organized as follows. First, in Chapter 2, we briefly review the literature of the problem. Then, in Chapter 3, we describe the proposal of this work. This paper continues with experimental results in Chapter 4 and conclusions in Chapter 5.

Chapter 2

Literature Review

2.1 Notion of Membership

Clustering is built on top of the notion of membership, which may be *Hard* or *Fuzzy*. In a model based on hard membership, each data item belongs to one cluster and is different from all other clusters. The fuzzy membership regime, however, maintains that each data item belongs to all clusters. K-means [4], Hard C-means (HCM) [5, 6], and Iterative Self-Organizing Data Clustering (ISODATA) [7] utilize hard membership values.

With the introduction of Fuzzy Theory, many researchers incorporated this more "natural" notion into clustering algorithms. The premise for employing a fuzzy clustering algorithm is that, generally, no distinct line of separation exists between the clusters [8]. Additionally, from a practical perspective, it is observed that hard clustering techniques are considerably more prone to falling into local minima [5] (also see [9, 10, 11]). The reader is referred to [12] for the wide array of fuzzy clustering methods developed in the past few decades and to [13] for a recent review of the field of fuzzy clustering.

Initial work on fuzzy clustering was done by Ruspini [14] and Dunn [15] and it was then generalized by Bezdek [16] into Fuzzy C-means (FCM). In FCM, data items, which are denoted as $x_1, \cdots, x_N$, belong to $\mathbb{R}^k$ and clusters, which are identified as $\psi_1, \cdots, \psi_C$, are represented as points in $\mathbb{R}^k$. FCM makes the assumption that the number of clusters, C, is known through a separate process or expert opinion and minimizes the following objective function,

$$\Delta = \sum_{c=1}^{C} \sum_{n=1}^{N} f_{nc}^m \|x_n - \psi_c\|^2. \tag{2.1}$$

This objective function is heuristically suggested to result in appropriate clustering results and is constrained by,

$$\sum_{c=1}^{C} f_{nc} = 1, \forall n. \tag{2.2}$$

Here, $f_{nc} \in [0, 1]$ denotes the membership of x_n to ψ_c.

In (2.2), $m > 1$ is the *fuzzifier* (also called *weighing exponent* and *fuzziness*). The optimal choice for the value of the fuzzifier is a debated matter [17] and is suggested to be "an open question" [18]. It has been suggested that $1 < m < 5$ [19] and $1.5 < m < 2.5$ [20] are proper ranges and that $m = 2$ is an appropriate value [19]. The use of $m = 2$ is suggested in early work on the topic [15, 21, 22] and there is physical evidence for this choice as well [11]. Nevertheless, other researchers [18] have argued that the choices for the value of m are mainly empirical and lack a theoretical basis. It is known that larger values of m soften the boundary between the clusters [23]. The reader is referred to [24] for a review of the concept of fuzzifier and the alternatives for it. In this work, we use $m = 2$ in order to comply with the history of the problem.

The success of FCM motivated many researchers to modify it in order to alter its behavior and to achieve more desirable properties. However, the augmentation of FCM with regularization terms and constraints is not without its own inherent hazards. In fact, one of the byproducts of that process is that the resulting clustering algorithm becomes dependent on additional parameters which need to be properly and carefully set. Often the "proper" setting of these parameters is critical for the function of the corresponding algorithms. For example, in [25], the authors state that the performance of their proposed method depends on five parameters which have to be "chosen through experience" and provide a list of some other affected approaches.

2.2 Prototype-based Clustering

It is a common assumption that the notion of homogeneity depends on the distances between the data items. This assumption is made implicitly when clusters are modeled as *prototypical* data items, also called *clustroids* or cluster *centroids*, as in FCM, for example. A prominent choice in these works is the use of the Euclidean distance function [26]. For example, the potential function approach considers data items as energy sources scattered in a multi-dimensional space and seeks peak values in the field [27] (also see [28, 29, 30]). We argue, however, that the *distance* between the data items may not be either defined or meaningful and what the clustering algorithm is to accomplish is the minimization of *data item-to-cluster* distances. For example, when data items are to be clustered into certain lower-dimensional subspaces, as it is the case with Fuzzy C-Varieties (FCV) [31], the Euclidean distance between the data items is *irrelevant*. We note that, in fact, fuzzy clustering is commonly equated to and reduced to prototype-based clustering [23]. This reductive perspective is prevalent as of 2015 [32]. In fact, a person not familiar with the field may easily conclude, based on the treatment of the problem in works such as Parallel Fuzzy C-Means (PFCM) [33], that fuzzy clustering is *inherently* and *exclusively* a prototype discovery mechanism. A similar reduction of fuzzy clustering to seeking prototypes is made in the design of the Dynamic Fuzzy Clustering (DFC) technique introduced in [34]. The collaborative fuzzy clustering algorithm proposed in [35] follows a similar perspective too. The reader is referred to [36] for a unified formalism of prototype-based clustering algorithms, what that paper calls the "CM Family", including FCM, HCM, Deterministic Annealing (DA) [29], and Possibilistic c-Means with an entropic cost term (PCM-II) [37].

Prototype-based clustering does not necessarily require prototypes which are explicitly *present*. For example, in kernel-based clustering [38], it is assumed that a non-Euclidean distance can be defined between any two data items. The clustering algorithm then functions based on an FCM-style objective function and produces clustroids which are defined in the same feature space as the data items [39]. These cluster prototypes may not be explicitly represented in the data item space, but, nevertheless, they share the same mathematical model as the data items [40]. Fuzzy Analysis (FANNY) [41] is another algorithm in which, although there are no prototypes, but, homogeneity is based on the mutual distances between the data items.

Relational clustering approaches constitute a generic class of algorithms which are *intrinsically* based on the pairwise distances between the data items (for example refer to Relational FCM (RFCM) [42] and its non-Euclidean extension Nerf C-means [43]). The goal of that class of algorithms is to group the data items into *self-similar* bunches. Another algorithm in which the presence of prototypes may be less evident is Multiple Prototype Fuzzy Clustering Model (FCMP) [44], in which data items are described as a linear combination of a set of prototypes, which are, nevertheless, members of the same $\mathbb{R}^k$ as the data items are. Fuzzy clustering by Local Approximation of Memberships (FLAME) [45] and Hierarchical Agglomerative Clustering (HAC) [46, 14.3.12 Hierarchical clustering] are other clustering algorithms which inherently guide the process of clustering based on the distances between the data items.

The use of prototype-based clustering algorithms leads to challenges specifically when complex notions of homogeneity are applicable to the problem class in hand. For example, in [25] the authors argue that multiple prototypes ought to be utilized when geometries other than spherical and ellipsoidal are to be addressed. They formalize that approach into the two-level Fuzzy Convex Clustering (FCC) algorithm, which is composed of consecutive FCC Expansion (FFCE) and FCC Merging (FFCM) stages. Through that mechanism, they employ convex polytopes, which they consider to represent "flexible prototypes". The reader is referred to [12] for examples of fuzzy clustering problems which involve non-Euclidean geometries.

We argue that a successful departure from the assumption of prototypical clustering is achieved when clusters and data items have different mathematical models. For example, the Gustafson-Kessel algorithm [47] models a cluster as a pair of a point and a covariance matrix and utilizes the Mahalanobis distance between data items and clusters (also see the Gath-Geva algorithm [48], Fuzzy C-Regression Models (FCRM) [49], and the improvements given in [50]). Fuzzy shell clustering algorithms [22], which are sometimes addressed as Fuzzy C-Shells (FCS), utilize more generic geometrical structures. For example, the FCV [31] algorithm can detect lines, planes, and other hyper-planar forms, the Fuzzy C Ellipsoidal Shells (FCES) [51] algorithm searches for ellipses, ellipsoids, and hyperellipsoids, and the Fuzzy C Quadric Shells (FCQS) [22] and its variants seek quadric and hyperquadric clusters. While these generalizations of the prototype-based model are very important, we note that they represent *uneconomical* and *individualistic* extensions. In other words, each one of the aforementioned algorithms generalizes FCM for one particular problem class, and, hence, they repeat both the mathematical derivation of the solution and its implementation as computer code.

2.3 Robustification

The function of the membership values in FCM and the concept of the weight functions in robust statistics are related [52]. In essence, the argument is that the classical FCM provides an indirect means for attempting robustness. Nevertheless, it is known that FCM and other least square methods are highly sensitive to noise [53]. A numerical review of this topic is carried in [54]. Hence, there has been ongoing research on the possible modifications of FCM in order to provide a (more) robust clustering algorithm [55, 56]. An extensive list of relevant works and an outline of the intrinsic similarities within a unified view can be found in [52] (also see [57, 58]).

The first attempt to robustifying FCM, based on one account [52], is the Ohashi Algorithm [57,

59]. That work adds a noise cluster to FCM and writes the robustified objective function as,

$$\Delta = \alpha \sum_{c=1}^{C} \sum_{n=1}^{N} f_{nc}^{m} \|x_n - \psi_c\|^2 + (1 - \alpha) \sum_{n=1}^{N} \left(1 - \sum_{c=1}^{C} f_{nc}\right)^m. \tag{2.3}$$

The transformation from (2.1) to (2.3) was suggested independently through the algorithm nicknamed Noise Clustering (NC) [58, 60] as well (also see Robust Fuzzy Clustering Algorithm (RFCA) [60]). The core idea in NC is that there exists one additional imaginary prototype which is at a fixed distance from all of the data items and represents noise.

NC was later extended into Possibilistic C-means (PCM) [61], wherein the cost function is rewritten as,

$$\Delta = \sum_{c=1}^{C} \sum_{n=1}^{N} t_{nc}^{m} \|x_n - \psi_c\|^2 + \sum_{c=1}^{C} \eta_c \sum_{n=1}^{N} (1 - t_{nc})^m. \tag{2.4}$$

Here, t_{nc} denotes the degree of representativeness or *typicality* of x_n to ψ_c (also addressed as a *possibilistic degree* in contrast to the *probabilistic* model utilized in FCM). As expected from the modification in the way t_{nc} is defined, compared to that of f_{nc}, PCM removes the sum of one constraint, shown in (2.2), and in effect extends the idea of one noise cluster in NC into C noise clusters. In other words, PCM could be considered as the parallel execution of C independent NC algorithms that each seek a cluster. Therefore, the value of C is somewhat arbitrary in PCM [52]. For this reason, PCM has been called a *mode-seeking* algorithm where C is the upper bound on the number of modes.

It is likely that PCM clusters coincide and/or leave out portions of the data unclustered [62]. In fact, it is argued that the fact that at least some of the clusters generated through PCM are non-coincidental is because PCM gets trapped into local minimum [63] (also see [24, 23]). PCM is also known to be more sensitive to initialization and the exact values of the configuration parameters than other algorithms in its class [62, 26].

The favorable and distinct features of FCM and PCM have motivated their combination into Fuzzy Possibilistic C-Means (FPCM) [64], which seeks the minimization of,

$$\Delta = \sum_{c=1}^{C} \sum_{n=1}^{N} (f_{nc}^{m} + t_{nc}^{\eta}) \|x_n - \psi_c\|^2, \tag{2.5}$$

subject to (2.2) and $\sum_{n=1}^{N} t_{nc} = 1, \forall c$. That approach was later shown to suffer from different scales for f_{nc} and t_{nc} values, especially when $N \gg C$, and, therefore, additional linear coefficients and a PCM-style term were introduced to the objective function [65]. It has been argued that the resulting objective function employs four correlated parameters and that the optimal choice for them for a particular problem instance may not be trivial [26]. Additionally, in the new combined form, f_{nc} cannot necessarily be interpreted as a membership value [26].

The introduction of the PCM model was motivated by several factors, amongst which is to be able to relax, or somewhat circumvent, the sum-of-one constraint for the membership values. As such, through "giving up the requirement for strict partitioning" [36], the expectation is that the resulting algorithm will be able to reject outliers and to deal with data items which do not belong to any of the clusters more efficiently. As discussed here, however, the utilization of the PCM-style models has given rise to the emergence of other difficulties. In this context, the model presented in [66] is worth particular attention. That work stated that the relationship between the data items and the clusters must be assessed at two levels, i.e. whether or not a data item is an outlier and, if not, which cluster(s) it belongs to. In other words, the model developed in [66] replaced the singleton membership identifier f_{nc} with the pair of p_n and f_{nc}. Here, p_n models the probability that x_n is an inlier and f_{nc} models the probability that it belongs to ψ_c, given that it is an inlier. An extension of this model in [67] demonstrated that the p_n variables can emerge from the f_{nc} values when the classical parallel clustering framework is converted to a serially-structured pipeline.

Another important contribution is Robust C-Prototypes (RCP) [21], which is based on embedding a robust loss function in the objective function of FCM, wherein the cost function is rewritten as,

$$\Delta = \sum_{c=1}^{C} \sum_{n=1}^{N} f_{nc}^{m} u_c \left(\|x_n - \psi_c\| \right). \tag{2.6}$$

Here, $u_c(\cdot)$ is the robust loss function for cluster c (also see Unsupervised RCP (UCRP) [21]). Alternative HCM (AHCM) and Alternative FCM (AFCM) [40] employ a similar framework and use $u_c(x) = 1 - e^{-\beta x^2}$.

2.4   Number of Clusters

The classical FCM and PCM, and many of their variants, are based on the assumption that the number of clusters is known (an extensive review of this topic is given in [16, Chapter 4]). While PCM-style formulations may appear to relax this requirement, the corresponding modification is carried out at the cost of yielding an ill-posed optimization problem [26]. An important 1997 paper on this topic concludes that the solution to the general problem of robust clustering, when the number of clusters is unknown, is "elusive" and that the techniques available in the literature each have their limitations [52].

Repeating the clustering procedure for different numbers of clusters [48] and *Progressive Clustering* are two of the approaches used to address the challenge of not requiring *a priori* knowledge about the number of clusters present in a particular set of data items. Among the many variants of Progressive Clustering are methods which start with a significantly large number of clusters and freeze "good" clusters [68, 69], approaches which combine compatible clusters [70, 68, 69, 21], and the technique of searching for one "good" cluster at a time until no more is found [71]. These approaches utilize one or more Cluster Validity Indexes (CVI) [72] in order to assess the appropriateness of the clusters produced after each execution of the algorithm. Use of regularization terms in order to push the clustering results towards the "appropriate" number of clusters is another approach taken in the literature [73]. These regularization terms, however, generally involve additional parameters which are to be set carefully, and potentially per problem instance [64].

As such, the thinking process behind some of the earliest attempts at the estimation of C, is one of "find many clusters, select only a few" [74]. In this line of thinking, the clustering algorithm is allowed to generate many cluster *candidates* and significant resources have been allocated to the study of cluster validity assessment techniques. The reader is referred to a comprehensive review of this after-the-fact approach in [75] (also see Dunn's index [15], the DB index [76], and the PBM index [77]). Nevertheless, a majority of those approaches make the assumption that the validity of a cluster, or an entire clustering solution, can be measured using a scalar value. However, thorough examination of 23 scalar measures of cluster validity, researchers have shown that "*none* of them are exceptionally reliable across a wide range of datasets" [78]. From a theoretical perspective, too, it has been argued that scalar cluster validity indexes *aggregate* the entire information available in an input set of data items into one or a few metrics and that invaluable information is lost in this process. In the words of the authors of [79], "scalar measures of cluster validity are famously unreliable".

The findings regarding the inherent deficiencies of scalar clustering validity indexes have encouraged the community to investigate alternative techniques which visualize or assess the inherent structure of a given set of data items from the vantage point of clustering. Visual Assessment of clustering Tendency (VAT) [80] utilizes the pairwise dissimilarity information between data items as a symmetrical matrix with non-negative elements and zero diagonals known as the Dissimilarity

Matrix (DM). VAT provides a mechanism for reordering the rows and columns of this matrix in a way that signifies the structure of the data. In short, when VAT is successful, dense areas in the dataset yield dark squares along the diagonal of the Reordered Dissimilarity Matrix (RDM). One can, therefore, at least theoretically, count these squares and generate a reasonable estimate of the number of clusters present in the data. We note that VAT must be reviewed in the context of other visualization techniques such as trees, dendrograms, castles, and icicles [81]. More specifically, VAT belongs to the subset of techniques which utilize image-based visualization mechanisms [82, 83].

It has been argued that "a major limitation" [84] of VAT and its variants is their "inability to highlight cluster structure ... when ... [the data] contains clusters with highly complex structure" [84]. Spectral VAT (SpecVAT) [84] attempts to increase the legibility of the RDM generated by VAT algorithms through spectral decomposition of the DM prior to the reordering [85]. However, the performance of SpecVAT depends on the proper selection of the parameter k, i.e. the number of eigenvectors used during decomposition. In fact, as demonstrated in [86], to properly select k, one ought to have a proper estimate for C, the number of clusters present in the data. The critical necessity for the existence of this *a priori* piece of information violates the premise behind VAT and SpecVAT.

In spite of their differences, VAT, Revised VAT (reVAT) [87], bigVAT [88], Scalable VAT (sVAT) [89], and many other algorithms in their class, are in fact *visual assessment* methods. In other words, these techniques are inherently reliant on the *subjective* [88] understanding of the user. This issue becomes more disconcerting when it is argued that in some practical settings, an "experienced user" [88] may have to be employed in order to perform the assessment. In fact, it has been argued that non-Euclidean geometries or overlap between the clusters can give rise to VAT images for which "different viewers may deduce different numbers of clusters ..., or worse, not be able to estimate c at all" [86]. To make matters more complicated, the output of many of the algorithms in this class is a two dimensional image, which requires to be transferred and displayed *diligently* and it is argued that compression, down-scaling, and interleaving of this image "may obscure important information about potential clusters in the data" [88].

For an input set of data items $\mathbf{X}$, which contains N data items, the computational complexity of VAT is of $O(N^2)$. This is due to the fact that VAT processes the $N \times N$ matrix of dissimilarities between the data items. In addition to the fact that this model is inappropriate in the context of a generic notion of homogeneity, the quadratic complexity of VAT is prohibitive for Big Data problem instances. In fact, it has been argued that VAT "works well for relatively small data sets

$(n \leq 500)$" [88]. Hence, variations of VAT have been proposed which address this issue. reVAT, bigVAT, and sVAT are three approaches which aim to lower the computational complexity to $O(CN)$. Here, C is the inherent, and *unknown*, number of clusters which are present in **X**. As commonly $C \ll N$, this transformation is greatly beneficial. Nevertheless, it has been argued that "sVAT does not asymptotically scale linearly with [the number of data items]" [90].

The $O(N^2)$ computational complexity of VAT contains the superpositions of the costs associated with two processes, both of which require $O(N^2)$ operations. In fact, not only the DM needs to be reordered at the cost of $O(N^2)$ operations, but also this matrix needs to be calculated in the first place, and the computational complexity of that process is of $O(N^2)$ as well. In this context, the sequels of VAT, i.e. reVAT, bigVAT, and sVAT, drop the complexity of the reordering mechanism to $O(CN)$, but, nevertheless, they still require the calculation of the $N \times N$ DM. Hence, technically, the computational complexities of those algorithms are still $O(N^2)$. Therefore, neither VAT nor the aforementioned variants of it are *genuinely* scalable. Here, we rely on the notion of scalability which mandates that the computational complexity of the algorithm must grow linearly as the number of input data items increases [91].

VAT approaches have been augmented with path-based distance models in order to alleviate the limiting scope mandated by the assumption of the relational model. For example, Improved VAT (iVAT) [86] prescribes that it is not the direct data item-to-data item distance which must be stated in the DM but that x_1 and x_2 are "similar" if there is a sequence of data items, with x_1 and x_2 at the two ends of the sequence, each of which are at close distances to each other [92]. Sample experimental results provided in [86] suggest that iVAT is capable of recognizing arbitrary sequences of data items in $\mathbb{R}^2$. The authors of that paper also propose the Automatic VAT (aVAT) technique which applies a function, that can be considered a robust loss function, on the RDM in order to facilitate the semi-unsupervised extraction of the diagonal blocks from it. A related algorithm, named clusiVAT [93], samples the input set of data items in order to generate the cluster representations using Single Linkage (SL) [94]. clusiVAT then extends the classification results to the entire set of data items. One scenario in which the path-based model falls short of resolving the challenge is the clustering of members of $\mathbb{R}^k$ into lower-dimensional spaces. In this scenario, any intersection between a pair of clusters is a real threat which causes the two clusters to "leak" into each other. This case is closely related to the "zigzagging" phenomenon which the original implementation of VAT was prone to and was alleviated using a clever initialization procedure for the reordering mechanism [80].

14

Chapter 3

Method

3.1 Modeling Framework

Any unsupervised data clustering problem is based on a *notion of homogeneity* which is rooted in the physical properties of the corresponding data items and clusters. Therefore, there is an important practical incentive for discussing unsupervised data clustering in generic terms and independent of the framework of any particular *problem class*. Here, a problem class is the mathematical formalization of the data clustering problem in the context of a particular model for the data items and a particular notion of homogeneity. This notion of homogeneity contains a cluster model. For example, one may refer to the problem class which relates to "Euclidean clustering of points which belong to $\mathbb{R}^2$". In this statement, "Euclidean clustering" defines the notion of homogeneity, which then mandates that each cluster is represented as a point in $\mathbb{R}^2$, thus defining the cluster model relevant to the problem class in hand. Additionally, the clause "points which belong to $\mathbb{R}^2$" provides the model for the data items. Once a problem class has been identified, one can discuss particular *problem instance*s. Here, a problem instance is one realization of a particular problem class. In other words, a problem instance provides a set of data items, as prescribed by a problem class.

Under such circumstances, the vision of this work is to develop a generic fuzzy clustering algorithm which accepts data item and cluster models as plug-ins and operates using a data item-to-cluster distance function which is provided as a black-box. In other words, we propose a generic unsupervised fuzzy clustering algorithm which can be adopted to any problem class. Once the adoption is carried out, this particular incarnation of the proposed method will operate on any instance of the aforementioned problem class without any need for user supervision or subjective intervention.

3.2 Model Preliminaries

We denote a data item as x and a cluster as ψ. Here, we assume that the problem instance in hand provides the weighted set of data items, defined as,

$$\mathbf{X} = \left\{ (\omega_n; x_n) \right\}, n = 1, \cdots, N, \omega_n > 0, \tag{3.1}$$

and we define the *weight* of $\mathbf{X}$ as,

$$\Omega(\mathbf{X}) = \sum_{n=1}^{N} \omega_n. \tag{3.2}$$

When known in the context, we abbreviate $\Omega(\mathbf{X})$ as Ω. Thus, when estimating expected values, we treat $\mathbf{X}$ as a set of realizations of the random variable x and write,

$$p\{x_n\} = \frac{\omega_n}{\Omega}. \tag{3.3}$$

We model the relationship between a data item and a cluster as the real-valued positive *distance function* $\phi(x, \psi)$. Through this abstraction, we decidedly avoid the dependence of the underlying algorithm on Euclidean or any other particular notations of distance.

We assume that the robust loss function, $u(\cdot) : [0, \infty] \to [0, 1]$, is given which satisfies $\lim_{\tau \to \infty} u(\tau) = 1$. Additionally, we assume that $u(\cdot)$ is an increasing differentiable function which satisfies $u(0) = 0$ and $u(1) = \frac{1}{2}$. In this work, we utilize the rational robust loss function,

$$u(x) = \frac{x}{1 + x}, \tag{3.4}$$

and we model the loss of x_n when it belongs to ψ_c, as,

$$u_{nc} = u\left(\frac{1}{\lambda}\phi_{nc}\right), \phi_{nc} = \phi(x_n, \psi_c). \tag{3.5}$$

We model the loss of a data item which is considered to be an outlier as the positive constant U. In (3.5), we address λ as the *scale* parameter (note the similarity with the cluster-specific weights in PCM [61]). In fact, λ has a similar role to that of scale in robust statistics (also called the *resolution parameter* [28]) and the idea of distance to noise prototype in the NC algorithm [58, 60]. Scale can also be considered as the controller of the boundary between inliers and outliers [52]. From a geometrical perspective, λ controls the radius of spherical clusters and the thickness of planar and shell clusters [26].

We assume that $\phi(x, \psi)$ is differentiable in terms of ψ and that for any non-empty weighted set $\mathbf{X}$, the following function of ψ,

$$\Delta_{\mathbf{X}}(\psi) = E\{\phi(x, \psi)\} = \frac{1}{\Omega} \sum_{n=1}^{N} \omega_n \phi(x_n, \psi), \tag{3.6}$$

has one and only one minimizer which is also the only solution to the following equation,

$$\sum_{n=1}^{N} \omega_n \frac{\partial}{\partial \psi} \phi(x_n, \psi) = 0. \tag{3.7}$$

In this paper, we assume that a function $\Psi(\cdot)$ is given, which, for the input weighted set $\mathbf{X}$, produces the optimal ψ which minimizes (3.6) and is the solution to (3.7). We address $\Psi(\cdot)$ as the *cluster fitting function.*

Note that $\Psi(\cdot)$ is the solution to the M-estimator given in (3.6). We emphasize that when a closed-form representation for $\Psi(\cdot)$ is not available, conversion to a W-estimator can produce a procedural solution to (3.7) [95]. Additionally, many of the techniques developed in the context of Weber Problems [96] may be applicable to finding a procedural solution to $\Psi(\cdot)$.

A version of this modeling framework has been called a "prototype generator" [79] in that a cluster is modeled using a finite set of mathematical entities which *generate* the set of data items that belong to the cluster. The entirety of this model has precedence in the literature and has been used in parallel [66] as well as sequential settings [67].

3.3   Assessment of Loss

In this section, we carry the loss modeling framework developed in [66] in order to render a self-containing paper. This framework derives a loss model for the set $\mathbf{X}$ which is *known* to contain C clusters. We then derive the single-cluster version of this loss model.

We assume that, at some arbitrary point during the procedure, a clustering algorithm has discovered the C clusters $\psi_1, \cdots, \psi_C$ in $\mathbf{X}$. We also assume that a Maximum Likelihood procedure has been applied on $\mathbf{X}$ and denote the set of data items which are assigned to ψ_c as $\tilde{\mathbf{X}}_c$. We address the union of all $\tilde{\mathbf{X}}_c$ for $c = 1, \cdots, C$ as $\tilde{\mathbf{X}}$. In this context, the set $\tilde{\mathbf{X}}_0 = \mathbf{X} - \tilde{\mathbf{X}}$ contains the data items which are considered to be outliers.

Now, we consider an arbitrary data item x_n. This data item may be an outlier or it may belong to one of the C clusters. Hence, we model the loss associated with x_n as follow.

$$E\{Loss|x_n\} = p\{x_n \in \tilde{\mathbf{X}}_0\} E\{Loss|x_n \in \tilde{\mathbf{X}}_0\} + \tag{3.8}$$

$$p\{x_n \in \tilde{\mathbf{X}}\} \sum_{c=1}^{C} p\{x_n \in \tilde{\mathbf{X}}_c|x_n \in \tilde{\mathbf{X}}\}$$

$$E\{Loss|x_n \in \tilde{\mathbf{X}}_c\}.$$

We now denote the probability that x_n is an inlier as p_n and the probability that x_n belongs to $\tilde{\mathbf{X}}_c$, given that it is an inlier, as f_{nc} and rewrite (3.8) as follows.

$$E\{Loss|x_n\} = (1 - p_n)U + p_n \sum_{c=1}^{C} f_{nc}u_{nc}, \tag{3.9}$$

where (3.5) is used. Note that the definition of f_{nc} mandates (2.2). Then, we aggregate (3.8) for all x_n and use (3.3) and write,

$$E\{Loss|\mathbf{X}\} = \sum_{n=1}^{N} p\{x_n\} E\{Loss|x_n\} = \tag{3.10}$$

$$\frac{1}{\Omega} \sum_{n=1}^{N} w_n \left[p_n \sum_{c=1}^{C} f_{nc}u_{nc} + UC\frac{1}{C}(1 - p_n) \right].$$

Close assessment of (3.10) shows that this cost function complies with an HCM-style hard template. It is known, however, as discussed in Section 2.1, that the utilization of the concept of the fuzzifier has important benefits. Hence, we incorporate m into (3.10) and derive the following cost function.

$$\Delta = \sum_{n=1}^{N} w_n \left[p_n^m \sum_{c=1}^{C} f_{nc}^m u_{nc} + UC^{1-m}(1 - p_n)^m \right]. \tag{3.11}$$

In (3.10) we have chosen to write U as $UC\frac{1}{C}$ in order to compensate for the impact of the fuzzifier. In fact, two types of terms exist in (3.10), i.e. $p_n f_{nc}$ and $(1 - p_n)$. These terms are each products of membership identifiers. However, while the first term contains two elements, the second one only contains the single element $1 - p_n$. We argue that this is because this term in fact contains implicit components of the type "*if either P or not P*" hidden in it. In other words, the cost component $U(1 - p_n)$ in fact models the situation in which x_n is an outlier, in which case it is irrelevant whether or not x_n belongs to any of the clusters. In other words, the term $(1 - p_n)$ is in fact the simplified version of the following term,

$$(1 - p_n) \sum_{c=1}^{C} f_{nc} = 1 - p_n. \tag{3.12}$$

While this alternative form is identical to $1 - p_n$, the difference between the two sides of (3.12) becomes significant when the fuzzifier is integrated into the objective function. In other words, with the addition of the fuzzifier, the term given in (3.12) ought to be modified to,

$$(1 - p_n)^m \sum_{c=1}^{C} f_{nc}^m \leq (1 - p_n)^m. \tag{3.13}$$

Hence, if no other measure is taken, the incorporation of the fuzzifier effectively reduces the cost of being an outlier, as explained below.

For any set of C non-negative variables ζ_c which satisfy $\sum_{c=1}^{C} \zeta_c = 1$, we have $\sum_{c=1}^{C} \zeta_c^m \leq 1$, when $m > 1$. The equality in this relationship, i.e. the upper bound, occurs when all of the ζ_c are zero except for one which is unity. The lower bound on $\sum_{c=1}^{C} \zeta_c^m$, however, occurs when the ζ_c are identical. Therefore, we replace (3.12) with the case in which all the f_{nc} are equal. This process guarantees that when the fuzzifier is incorporated into the cost function, the corresponding term is always greater than or equal to the pre-fuzzifier term. In other words, we replace $(1 - p_n)$ with $(1 - p_n)C\frac{1}{C}$ and, therefore, after the incorporation of the fuzzifier, yield $(1 - p_n)^m C\frac{1}{C^m} = (1 - p_n)^m C^{1-m}$. In the above, this transformation was rephrased, *imprecisely*, as substituting U with $UC\frac{1}{C}$ in (3.10).

3.4   Determination of U and λ

The notion that a fuzzy clustering algorithm functions within an *unsupervised* framework is often misunderstood and inaccurately stated. In effect, unsupervised clustering is commonly defined in contrast to *supervised* clustering, wherein a *supervisor* provides training data and possibly direction and guidance in order for the algorithm to function desirably. Unsupervised clustering, on the other hand, is the term used for the broad category of algorithms which *theoretically* do not require user supervision. Nevertheless, as pointed out in reference to numerous pieces of work in the literature in Chapter 2, many available "unsupervised" clustering algorithms depend on the diligent adjustment of one or several configuration parameters. In many cases, these parameters are to be adjusted using trial-and-error or "user knowledge". We emphasize, however, that the only type of configuration variables acceptable to be employed by a *truly* unsupervised algorithm are those for which a deterministic and objective selection procedure is outlined which allows the user to be agnostic of the internal mechanics of the algorithm under consideration.

The execution of the algorithm developed in this paper is governed through the two configuration

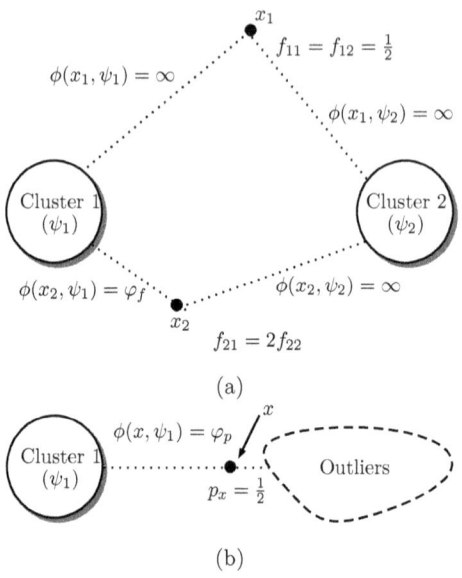

Figure 3.1: The process of determining λ and U. (a) Selection of φ_f. (b) Selection of φ_p.

parameters λ and U. In this section, we provide objective procedures for determining the values of these two configuration variables for any arbitrary problem class. It is worth emphasizing that λ and U are to be set at the level of problem classes and that they are not affected by the particular set of input data items associated with any problem instance or the number of clusters present in it.

It is evident that λ defines the scale for ϕ_{nc}. This is exemplified in (3.5) and also everywhere else in this paper where ϕ_{nc} is divided by λ. We argue that, similarly, U defines the scale for u_{nc} (for example see (3.21), and more specifically (3.20)). Hence, we argue that the two identities ϕ_{nc} and u_{nc} are *brought into context* through λ and U, respectively. We use this perceptual definition in order to propose procedures for determining the appropriate values for λ and U for an arbitrary problem class. We emphasize that in this process we utilize the multiple-cluster cost function carried in (3.11).

We suggest an imaginary situation, as depicted in Figure 3.1(a), in which two data items interact with two clusters ($N = 2$ and $C = 2$). The first data item, here x_1, is infinitely far from both clusters, in which case we expect $f_{11} = f_{12} = \frac{1}{2}$. The second data item, here x_2, however, is infinitely far from the second cluster and is at the distance of φ_f from the first one. Here, we ask what value of φ_f will result in $f_{21} = 2f_{22}$. This question can be asked in a different setting as follows: *For a data item which is infinitely far from two clusters, how close should it get to one cluster, while maintaining its distance to the other, in order for it to be favored by the former cluster two times than the latter?* One can show that setting $m = 2$ in the general case given in (3.11), we have [66],

$$f_{nc} = \frac{u_{nc}^{-1}}{\sum\limits_{c'=1}^{C} u_{nc'}^{-1}}, \qquad (3.14)$$

based on which,

$$\lambda = \frac{\varphi_f}{u^{-1}\left(\dfrac{1}{2}\right)} = \varphi_f. \qquad (3.15)$$

In order to estimate the proper value for U, we utilize another imaginary situation in which one data item interacts with one cluster, as depicted in Figure 3.1(b). Here, we ask how far the data item should be from the cluster in order for it to be an outlier with a probability of half. This situation, in effect, defines the boundary of inliers and outliers when Maximum Likelihood is applied on p_n. We denote this distance as φ_p and utilize the fact that setting $m = 2$ in the general case given in (3.11) yields [66],

$$p_n = \frac{\sum\limits_{c=1}^{C} u_{nc}^{-1}}{C\dfrac{1}{U} + \sum\limits_{c=1}^{C} u_{nc}^{-1}}, \qquad (3.16)$$

and therefire,

$$U = u\left(\frac{1}{\lambda}\varphi_p\right). \qquad (3.17)$$

The two entities φ_p and φ_f reflect different aspects of the relationship between data items and clusters. Nevertheless, conceptually, we can envision that one may want to set $\varphi_p = \varphi_f = \varphi_o$. In this line of reasoning, φ_o denotes the territory of a cluster, within which the cluster considers a data item as an inlier, therefore $p_x \geq \frac{1}{2}$, and also owns the data item when in competition with another farther cluster, therefore $f_{xc} \geq \frac{1}{2}$. Reworking (3.15) and (3.17) for $\varphi_p = \varphi_f = \varphi_o$ we arrive at $\lambda = \varphi_o$ and $U = \frac{1}{2}$.

We conjecture that an alternative mechanism for determining λ and U is to infer these identities based on a number of homogeneous subsets which are provided as training data.

3.5 Single-Cluster Clustering

In this section, we set $C = 1$ and derive the single-cluster version of the loss model which was developed in Section 3.3. We then provide a solution strategy for it.

When $C = 1$, the constraint (2.2) reduces the set $\{f_{n1}, \cdots, f_{nC}\}$ to $\{1\}$. Therefore, we drop the subscript c from u_{nc} and rewrite (3.11) as,

$$\Delta = \sum_{n=1}^{N} \omega_n \left[p_n^2 u_n + (1 - p_n)^2 U \right]. \tag{3.18}$$

Note that Δ is to be minimized independent of any constraint.

Calculating $\frac{\partial \Delta}{\partial p_n}$ and equating it to zero, we derive the optimal p_n as,

$$p_n = \frac{U}{U + u_n}. \tag{3.19}$$

It is informative to rewrite (3.19) as,

$$p_n = 1 - u \left(\frac{1}{U} u_n \right). \tag{3.20}$$

Here, we observe that p_n in fact inversely depends on the value of the robust loss function $u(\cdot)$, when applied on u_n and wherein U is the scale. This provides further insight into why U is addressed as scale for u_n in Section 3.4.

One can show that (3.19) yields,

$$p_n = \frac{1 + \frac{1}{\lambda} \phi_n}{1 + \frac{1}{\lambda} \left(1 + \frac{1}{U} \right) \phi_n}. \tag{3.21}$$

For x_n to be associated with ψ, within a Maximum Likelihood framework, one would require $p_n \geq \frac{1}{2}$. Routine derivation shows that this condition, given (3.21), translates into,

$$\phi_n \leq \frac{1}{\frac{1}{U} - 1} \lambda. \tag{3.22}$$

We then derive,

$$\frac{\partial \Delta}{\partial \psi} = \sum_{n=1}^{N} \omega_n p_n^2 \frac{1}{\lambda} u' \left(\frac{1}{\lambda} \phi_n \right) \frac{\partial}{\partial \psi} \phi(x_n, \psi). \tag{3.23}$$

Using (3.7) we know that the solution to (3.23) is given as,

$$\psi = \Psi \left(\left\{ (\tilde{\omega}_n; x_n) \right\} \right). \tag{3.24}$$

Here,

$$\tilde{\omega}_n = \frac{\omega_n p_n^2}{\left(1 + \frac{1}{\lambda}\phi_n\right)^2} = \frac{\omega_n}{\left(1 + \frac{1}{\lambda}\left(1 + \frac{1}{U}\right)\phi_n\right)^2}. \tag{3.25}$$

Now, we plug (3.21) in (3.18) and write,

$$\Delta = \sum_{n=1}^{N} \omega_n \frac{\frac{1}{\lambda}\phi_n}{1 + \frac{1}{\lambda}\left(1 + \frac{1}{U}\right)\phi_n}. \tag{3.26}$$

Derivation shows that (3.26) can be rewritten as follows, while it is important to emphasize that the right side of (3.27) is to be *maximized*.

$$\Delta \equiv \sum_{n=1}^{N} \frac{\omega_n}{1 + \frac{1}{\lambda}\left(1 + \frac{1}{U}\right)\phi_n}. \tag{3.27}$$

As will be shown, the algorithm developed in this paper does not in fact need to calculate p_n. In fact, the entities which need to be calculated, within a Picard iteration, are $\tilde{\omega}_n$, Δ, and c_n. Here, c_n is one iff $p_n \geq \frac{1}{2}$. We now show that one can rewrite (3.22), (3.25), and (3.27) using the substitute variable $\bar{\phi}_n$, as defined below,

$$\bar{\phi}_n = 1 + \left(1 + \frac{1}{U}\right)\frac{1}{\lambda}\phi_n = 1 + \frac{3}{\lambda}\phi_n, \tag{3.28}$$

Here, we have used $U = \frac{1}{2}$, as shown in Section 3.4. Routine derivation shows that,

$$c_n = \left[\bar{\phi}_n \leq \frac{2}{1 - U}\right] = \left[\bar{\phi}_n \leq 4\right]. \tag{3.29}$$

Here, $[p]$ is the *Iverson bracket*, wherein $[p]$ is one(zero) if the Boolean variable p is true(false). Substituting (3.28) in (3.25) yields,

$$\tilde{\omega}_n = \frac{\omega_n}{\bar{\phi}_n^2}, \tag{3.30}$$

and substituting (3.28) in (3.27) results in,

$$\Delta \equiv \sum_{n=1}^{N} \frac{\omega_n}{\bar{\phi}_n}. \tag{3.31}$$

Here, we carry a brief overview of the single-cluster clustering procedure developed in this paper.

- **Inputs**

(a) Cluster representation ψ.

(b) Input set of data items $\mathbf{X}$.

- **Outputs**

  (a) Updated cluster representation ψ.

  (b) Classification identifiers $c_n, n = 1, \cdots, N$.

- **Procedure**

  (a) Loop

      (a) Calculate $\bar{\phi}_n$ for $n = 1, \cdots, N$, using (3.28).

      (b) Calculate $\tilde{\omega}_n$ for $n = 1, \cdots, N$, using (3.30).

      (c) Calculate ψ using (3.24).

      (d) Calculate Δ using (3.31).

      (e) If change in Δ is negligible, break the loop.

  (b) Calculate c_n for $n = 1, \cdots, N$, using (3.29).

3.6 Cluster Space Sampling

The process carried at the end of Section 3.5 inputs a cluster representation and modifies it into a *more optimal* representation. In effect, this process accepts a point in the cluster space and updates it through a local search mechanism. The fact that that process functions locally is an important asset, because it can ignore the rest of the data items and only "focus" on a locality of $\mathbf{X}$. Nevertheless, this same phenomenon means that it is probable that ψ would in fact correspond to a local minimum of the search space and that there is no guarantee that in every execution of that process, using any input ψ, the generated cluster in fact describes one of the major homogenous sets present in the input set of data items.

Additionally, it is important to discuss the mechanism through which ψ is generated. Here, we denote three potential sources for ψ, i.e. cluster space sweeps, *a priori* pieces of information, and random cluster representations. We describe these three mechanisms in detail in the next paragraphs. Each one of these processes generates a cluster population, which we denote as $\hat{\Psi}$. Note that, the number of elements in this set is independent of C and is determined based on the available budget for processing power and also model complexity.

First, $\hat{\mathbf{\Psi}}$ may be generated by a process which "sweeps" the cluster space, given that the cluster model allows for such an action. For example, in one imaginary situation, the cluster representation may contain a point in a rectangular subset, for example $[-L, L] \times [-L, L]$, and an angle. In such circumstances, having made a decision about the size of $\hat{\mathbf{\Psi}}$, one can devise step sizes along the three dimensions of the cluster space and then to uniformly sample this space. The sweeping mechanism provides assurance that the entire cluster space is examined. Nevertheless, designing a sweeping mechanism for an arbitrary cluster representation may be non-trivial and cumbersome.

A second approach to producing $\hat{\mathbf{\Psi}}$ is to utilize *a priori* pieces of information. For example, due to the inherent properties of the problem class, it may be known that clusters tend to agglomerate in certain "areas" in the cluster space. If this condition is applicable to a problem class, then one may capitalize on the opportunity and generate $\hat{\mathbf{\Psi}}$ as the concatenation of cluster representations located in these "hot spots".

The third approach, which is also the one adopted in this paper, is to populate $\hat{\mathbf{\Psi}}$ with random cluster representations which are generated using i.i.d. processes. This approach is easy to implement, especially when ψ can be represented as a vector in a hyper-rectangular subspace of some $\mathbb{R}^k$.

We assume that one of the above mechanisms, or any other applicable mechanism, has been employed and that the set $\hat{\mathbf{\Psi}}$, which contains $\hat{C}$ cluster representations is generated. Here, $\hat{C}$ is significantly larger than any estimate for C. We feed these cluster representations to the algorithm described in Section 3.5 and generate the updated set $\check{\mathbf{\Psi}}$ as well as the corresponding p_n elements for every data item and every member of $\check{\mathbf{\Psi}}$. Note that it is possible that $\check{\mathbf{\Psi}}$ may in fact contain less than $\hat{C}$ elements, because there is the possibility that the process outlined at the end of Section 3.5 may in fact be followed by a *validation* stage which may find some ψs *undesirable*, given the particularities of the problem class at hand. We note the number of elements of $\check{\mathbf{\Psi}}$ as $\check{C}$. In this work, we employ a *cluster weight* condition for this purpose. As such, ψ is only accepted if the weight of ψ is greater than θ_Ω, where $\theta_\Omega = 0.1$. One can show that the weight of ψ can be calculated as,

$$\Omega_\psi = E\left\{[x \in \tilde{\mathbf{X}}_\psi]\right\}. \tag{3.32}$$

The computational complexity of the process outlined in this section is of $O(N\hat{C})$, wherein $\hat{C}$ is independent of both C and N. In other words, the computational complexity of the method developed in this paper scales linearly with the size of the input set of data items and, hence, this method is *scalable* [91]. We will analyze this matter more carefully in Chapter 4.

3.7   Cluster Aggregation

The processes described in Sections 3.5 and 3.6 collectively generate a set of *locally optimal* clusters corresponding to the input set of data items $\mathbf{X}$. We emphasize that, here, "optimal" is defined in an exclusively local context. In other words, the members of $\check{\mathbf{\Psi}}$ are generated independent of each other. This is in direct contrast to FCM and the other algorithms in its class, wherein the C clusters, which are generated by the algorithm, are in fact produced simultaneously and by an intimately interlocked mechanism. In other words, if FCM or another one of its many variants are executed two times, once to generate C clusters and again to generate $C+1$ clusters, the two sets of clusters are generally inherently different, because every cluster in the second execution is influenced by the presence of the other C clusters. As will be shown later, the method developed in this work functions differently due to the fact that it generates the candidate clusters independently. Also, in this work, the size of $\hat{\mathbf{\Psi}}$ is determined merely by the processing budget allotted to the algorithm. Hence, it is not *unreasonable* to expect that once *enough* resources are spent on producing $\check{\mathbf{\Psi}}$, then this set will contain the dominant clusters present in $\mathbf{X}$. We will return to this point later in this paper.

In this work, we examine $\check{\mathbf{\Psi}}$ and determine its *unique* members. This process generates the summary set $\bar{\mathbf{\Psi}}$, members of which are drawn from $\check{\mathbf{\Psi}}$. Note that every member of $\bar{\mathbf{\Psi}}$ is guaranteed to meet a set of minimum requirements and it is then up to the user to decide how many of the members of $\bar{\mathbf{\Psi}}$ they would like to pick. In other words, we argue that, any reference to the size of $\bar{\mathbf{\Psi}}$, as if it is an entity which is "out there" is *bogus*, because no input set of data items, which corresponds to some actual physical phenomenon, contains 10, or 3 or 7 clusters. In other words, the question is *not* "how many clusters there are in $\mathbf{X}$". Instead, the algorithm must be able to produce a set of clusters and to guarantee that if a smaller set of clusters is to be returned, then the latter set will mostly contain the more prominent members of the former set. In the opposite direction, as well, a larger instance of $\bar{\mathbf{\Psi}}$, for the same problem instance, must mostly augment the more prominent clusters with less prominent ones. In other words, for $\bar{C}_1 < \bar{C}_2$, if one generates the corresponding sets of clusters $\bar{\mathbf{\Psi}}_{\bar{C}_1}$ and $\bar{\mathbf{\Psi}}_{\bar{C}_2}$, it must be highly likely that many of the members of $\bar{\mathbf{\Psi}}_{\bar{C}_1}$ have a counterpart in $\bar{\mathbf{\Psi}}_{\bar{C}_2}$. We emphasize that neither FCM, nor PCM or any other of their many variants, can provide this functionality.

Comparison of two clusters, when a generic cluster model is taken into consideration, is a challenging task. One approach to this problem contains comparison strategies which are inherently and deeply rooted in particular problem classes. For example, one many refer to the distance between points/lines/planes or differences in rotation angle/area/perimeter, or other similar geometrical entities. Aside from these overtly special cases, one generic approach to cluster comparison follows the idea utilized in Visual Cluster Validity (VCV) [79], wherein the model parameters corresponding to each cluster are "lumped" into a vector and the two vectors corresponding to the two clusters are compared. That approach is trivial to implement and yet lacks theoretical justification, because, for example, in the context of an ellipsoidal geometry, VCV will combine the components of the mean vector with the elements of the covariance matrix. We argue that that approach mixes into a pot entities which not only accept values at very different scales, but also, and more importantly, are different from a theoretical perspective.

In this work we utilize the N-element vector $\mathbf{p}_{\check{c}} = \{p_{\check{c}1}, \cdots p_{\check{c}N}\}$, for $1 \leq \check{c} \leq \check{C}$, as the *signature* for the cluster $\psi_{\check{c}} \in \check{\Psi}$. Note that, $\mathbf{p}_{\check{c}} \in [0,1]^N$, but $\mathbf{1}^T \mathbf{p}_{\check{c}} \neq 1$. In other words, $\mathbf{p}_{\check{c}}$ is not a probability distribution. We use the Iverson bracket notation and denote by $[\mathbf{p}_{\check{c}}]$ the crisp version of $\mathbf{p}_{\check{c}}$, and by $[\sim \mathbf{p}_{\check{c}}]$ its inverse. In this context, we propose that the two clusters $\psi_{\check{c}_1}$ $\psi_{\check{c}_2}$ are similar when $[\mathbf{p}_{\check{c}_1}]$ and $[\mathbf{p}_{\check{c}_2}]$ denote that the two clusters claim membership to the same set of data items. For example, we rewrite (3.32) using the notion of cluster signatures as,

$$\Omega_{\psi_{\check{c}}} = [\mathbf{p}_{\check{c}}]^T \mathbf{\Omega} [\mathbf{p}_{\check{c}}] = \frac{1}{\Omega} \sum_{\check{\phi}_{n\check{c}} \leq 4} \omega_n. \tag{3.33}$$

Here, $\mathbf{\Omega}$ is a diagonal matrix with ω_n as its elements, wherein the matrix is divided by Ω so that its trace is one.

We employ the $[\mathbf{p}_{\check{c}_1}]$ vectors and utilize a Bayesian risk-based comparison framework. In other words, we suggest that the two cluster representations $\psi_{\check{c}_1}$ and $\psi_{\check{c}_2}$, are *dissimilar* to the extent that they produce contradicting classification results. To do this, we utilize the two probabilities $E\left\{p\{x \notin \mathbf{X}(\psi_{\check{c}_2}) | x \in \mathbf{X}(\psi_{\check{c}_1})\}\right\}$ and $E\left\{p\{x \notin \mathbf{X}(\psi_{\check{c}_1}) | x \in \mathbf{X}(\psi_{\check{c}_2})\}\right\}$. For convenience, we address these two expected probabilities as $E\left\{-\psi_{\check{c}_2} | \psi_{\check{c}_1}\right\}$ and $E\left\{-\psi_{\check{c}_1} | \psi_{\check{c}_2}\right\}$, respectively, and define,

$$\delta_{\psi_{\check{c}_1}, \psi_{\check{c}_2}} = \max\left(E\left\{-\psi_{\check{c}_2} | \psi_{\check{c}_1}\right\}, E\left\{-\psi_{\check{c}_1} | \psi_{\check{c}_2}\right\}\right). \tag{3.34}$$

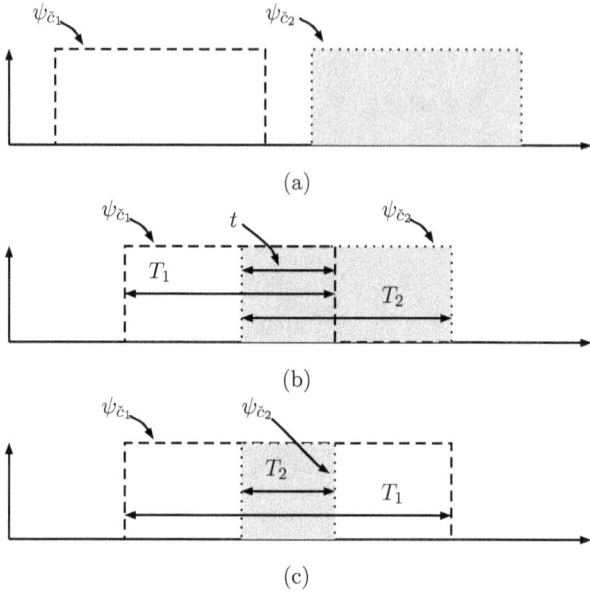

Figure 3.2: Symbolic representations of three pairs of clusters. (a) Distinct clusters. (b) Intersecting clusters. (c) Encompassing cluster.

Following the same logic utilized for the derivation of (3.32), we write,

$$E\left\{-\psi_{\check{c}_2}|\psi_{\check{c}_1}\right\} = \frac{E\left\{[x \in \tilde{\mathbf{X}}_{\psi_{\check{c}_1}} - \tilde{\mathbf{X}}_{\psi_{\check{c}_2}}]\right\}}{E\left\{[x \in \tilde{\mathbf{X}}_{\psi_{\check{c}_1}}]\right\}} = \frac{[\sim \mathbf{p}_{\check{c}_2}]^T \mathbf{\Omega}[\mathbf{p}_{\check{c}_1}]}{[\mathbf{p}_{\check{c}_1}]^T \mathbf{\Omega}[\mathbf{p}_{\check{c}_1}]} = \frac{\displaystyle\sum_{\bar{\phi}_{n\check{c}_1} \leq 4, \bar{\phi}_{n\check{c}_2} > 4} \omega_n}{\displaystyle\sum_{\bar{\phi}_{n\check{c}_1} \leq 4} \omega_n}. \qquad (3.35)$$

We note that one may utilize the Jaccard distance or other appropriate probabilistic distance metrics [97] instead of the model carried in (3.34).

Figure 3.2 provides symbolic representations of three pairs of cluster signatures. Here, we review these cases. In Figure 3.2, the horizontal axis denotes n, the indexes of the data items, and the vertical axis spans from zero to one and denotes crisp membership. Hence, the graphs denote $[\mathbf{p}_{\check{c}_1}]$ and $[\mathbf{p}_{\check{c}_2}]$.

In the first case, shown in Figure 3.2(a), the two clusters are distinct. Examination shows that in this case (3.35) yields that $E\left\{-\psi_{\check{c}_2}|\psi_{\check{c}_1}\right\}$ and $E\left\{-\psi_{\check{c}_1}|\psi_{\check{c}_2}\right\}$ are one, and hence, based on (3.34), $\delta_{\psi_{\check{c}_1},\psi_{\check{c}_2}}$ is one. In other words, these two clusters are completely distinct, which is what is also observed in Figure 3.2(a).

The case shown in Figure 3.2(b) denotes two clusters which intersect. In this case, (3.35) yields that $E\left\{-\psi_{\bar{c}_2}|\psi_{\bar{c}_1}\right\} = \frac{T_1-t}{T_1}$ and $E\left\{-\psi_{\bar{c}_1}|\psi_{\bar{c}_2}\right\} = \frac{T_2-t}{T_2}$. Hence, based on (3.34),

$$\delta_{\psi_{\bar{c}_1},\psi_{\bar{c}_2}} = 1 - \frac{t}{\max\left(T_1,T_2\right)}. \tag{3.36}$$

The significance of (3.35) becomes more evident when we analyze the case shown in Figure 3.2(c). In this case, effectively, $T_2 = t$. Hence, although $E\left\{-\psi_{\bar{c}_1}|\psi_{\bar{c}_2}\right\}$ is zero, $E\left\{-\psi_{\bar{c}_2}|\psi_{\bar{c}_1}\right\}$ is non-zero, and, therefore, this term holds $\delta_{\psi_{\bar{c}_1},\psi_{\bar{c}_2}}$ high. This is important, because otherwise a small, potential artifact of the local search, could claim similarity with two distinct clusters and thus push them into the same category.

Having defined $\Omega_{\psi_{\bar{c}}}$ and $\delta_{\psi_{\bar{c}_1},\psi_{\bar{c}_2}}$, we now propose a procedure which aggregates $\check{\mathbf{\Psi}}$ into $\bar{\mathbf{\Psi}}$. Here, $\bar{\mathbf{\Psi}}$ is the set of clusters reported by the method developed in this work. This process is governed by the two thresholds θ_δ, i.e. maximum cluster dissimilarity, and θ_τ, i.e. minimum cluster prominence. We define the notion of *cluster prominence* later in this section.

We represent any member of $\bar{\mathbf{\Psi}}$, using its cluster representation $\psi_{\bar{c}}$ accompanied by its weight, i.e. $\omega_{\psi_{\bar{c}}}$, as well as its prominence, i.e. $\tau_{\psi_{\bar{c}}}$. Both of these metrics belong to the interval $[0,1]$ and for any set $\bar{\mathbf{\Psi}}$, $\sum_{\psi_{\bar{c}} \in \bar{\mathbf{\Psi}}} \omega_{\psi_{\bar{c}}} \leq 1$ and $\sum_{\psi_{\bar{c}} \in \bar{\mathbf{\Psi}}} \tau_{\psi_{\bar{c}}} \leq 1$. Note that we have differentiated the weight of cluster $\psi_{\bar{c}}$, i.e. $\omega_{\psi_{\bar{c}}}$, when it is discussed in the context of a set of clusters, i.e. $\bar{\mathbf{\Psi}}$, from its independent weight, i.e. $\Omega_{\psi_{\bar{c}}}$, which is calculated based on (3.33), by using different Greek letters. This distinction is utmost important, because $\Omega_{\psi_{\bar{c}}}$ measures the portion of $\mathbf{X}$ which are classified into $\psi_{\bar{c}}$ when there is a dichotomy between $\psi_{\bar{c}}$ and the set of outliers. The former, i.e. $\omega_{\psi_{\bar{c}}}$, however, measures the share of $\psi_{\bar{c}}$ from $\mathbf{X}$ when the stakeholders are not only $\psi_{\bar{c}}$ and the set of outliers, but also the entire $\bar{\mathbf{\Psi}}$. In this context, a data item x_n belongs to $\tilde{\mathbf{X}}[\psi_{\bar{c}}]$ when, not only $p_{\bar{c}n} \geq \frac{1}{2}$, but also $p_{\bar{c}n}$ is greater than $p_{\bar{c}'n}$ for any $1 \leq \bar{c}' \leq \bar{C}$ and $\bar{c}' \neq \bar{c}$. One can show that,

$$p_{\bar{c}_1n} \geq p_{\bar{c}_2n} \leftrightarrow \phi_{\bar{c}_1n} \leq \phi_{\bar{c}_2n} \leftrightarrow \bar{\phi}_{\bar{c}_1n} \leq \bar{\phi}_{\bar{c}_2n}. \tag{3.37}$$

In other words, x_n is allotted to $\tilde{\mathbf{X}}[\bar{c}]$, when $\bar{c}$ satisfies not only that $\phi_{\bar{c}n} \leq 4$, but also that $\phi_{\bar{c}n}$ is the smallest member of $\{\phi_{1n},\cdots,\phi_{\bar{C}n}\}$.

We now formally describe the aggregation process utilized in this work.

- **Inputs**

  (a) Set of locally optimal clusters $\check{\mathbf{\Psi}} = \{(\psi_{\check{c}};[\mathbf{p}_{\check{c}}])\}$.

  (b) Maximum cluster dissimilarity θ_δ.

    (c) Minimum cluster prominence θ_τ.

- **Outputs**

    (a) Output set of clusters $\bar{\Psi} = \{(\psi_{\bar{c}}; \omega_{\psi_{\bar{c}}}, \tau_{\psi_{\bar{c}}})\}$.

- **Procedure**

    (a) Set $\bar{\Psi} = \varnothing$.

    (b) For every $\psi_{\check{c}} \in \check{\Psi}$,

        (a) For every $\psi_{\bar{c}} \in \bar{\Psi}$,

            i. If $\delta_{\psi_{\check{c}}, \psi_{\bar{c}}} \leq \theta_\delta$, then increment $\tau_{\psi_{\bar{c}}}$ and go to the next $\check{c}$.

        (b) Add $\psi_{\check{c}}$ to $\bar{\Psi}$ accompanied with $\tau_{\psi_{\bar{c}}} = 1$.

    (c) Normalize $\tau_{\psi_{\bar{c}}}$ for every member of $\bar{\Psi}$ by dividing it by $\check{C}$.

    (d) For every $\psi_{\bar{c}} \in \bar{\Psi}$,

        (a) Calculate $\omega_{\psi_{\bar{c}}}$ as the number of $x_n \in \mathbf{X}$, for which $\phi_{n\bar{c}} \leq 4$ and $\phi_{n\bar{c}}$ is the smallest member of $\{\phi_{1n}, \cdots, \phi_{\bar{C}n}\}$, divided by N.

It is important to point out that the computational complexity of this process can be estimated as $O(N\hat{C}\bar{C})$. In other words, the process developed in this work scales linearly with the number of data items present in the problem instance, as well as with the number of candidate clusters which are examined and the number of clusters which are presented as output. This linear cost system is of important significance when one contemplates Big Data problems in which $O(N^2)$ is fiercely non-practical and wherein $O(\hat{C}^2)$ is a serious challenge, because it in effect implies that the system is only capable of examining a large number of cluster candidates if processing budget can be extended on a hyperbolic curve. We will discuss the computational complexity of the developed method in more details later in this paper.

Note that, in this work, we determine if clusters are *similar* and then include them in the same set of clusters. One may consider a more elaborate aggregation scheme in which a set of similar clusters are used collectively in order to generate a cluster representation which is more optimal than each of them for the common set of data items that they lay claim to. In other words, one may replace the present take-first-instance framework with a use-all-instances structure.

Chapter 4

Experimental Results

The algorithm developed in this work is implemented as the Matlab class *Donna*. This class contains the core operations defined in this work, which, as stated, are class-independent. The child classes of this class override the two virtual functions $\phi(\cdot)$ and $\Psi(\cdot)$ and set the value of λ according to the specifications of their corresponding problem classes and the procedure outlined in Section 3.4. These child classes also perform the load and visualization operations relevant to the cluster and data item models defined by the respective problem classes. In this work, we utilize six such child classes, as listed in Table 4.1.

The process outlined in this paper starts with a problem instance and executes a cluster discovery mechanism through multiple repetitions of a local search supplemented with an aggregation mechanism. Hence, the "state" of the present algorithm can be examined at different stages and from different perspectives. Here, we utilize a sample 2dl problem instance and demonstrate some of the ways through which this problem instance, the operation of the proposed algorithm on it, and the solution generated by the proposed algorithm for it can be visualized.

Figure 4.1(a) carries a sample 2dl problem instance, wherein $\mathbf{X}$ contains 481 data items. This problem instance is processed using the developed algorithm and, after $3,250$ milliseconds of operation, the 3 clusters visualized in Figure 4.1(e) are generated. In the following paragraphs, we review the different aspects of this execution which are carried in Figure 4.1.

Figures 4.1(b), (c), and (d) show the individual members of $\bar{\Psi}$. These are in fact the distinct members of $\check{\Psi}$ which are, first, dissimilar enough from each other and, second, every other member of $\check{\Psi}$ is similar to one of them. Figure 4.1(e) shows the super-position of these 3 clusters into the same axis system. In this experiment, we used a population of $\hat{C} = 255$ cluster candidates.

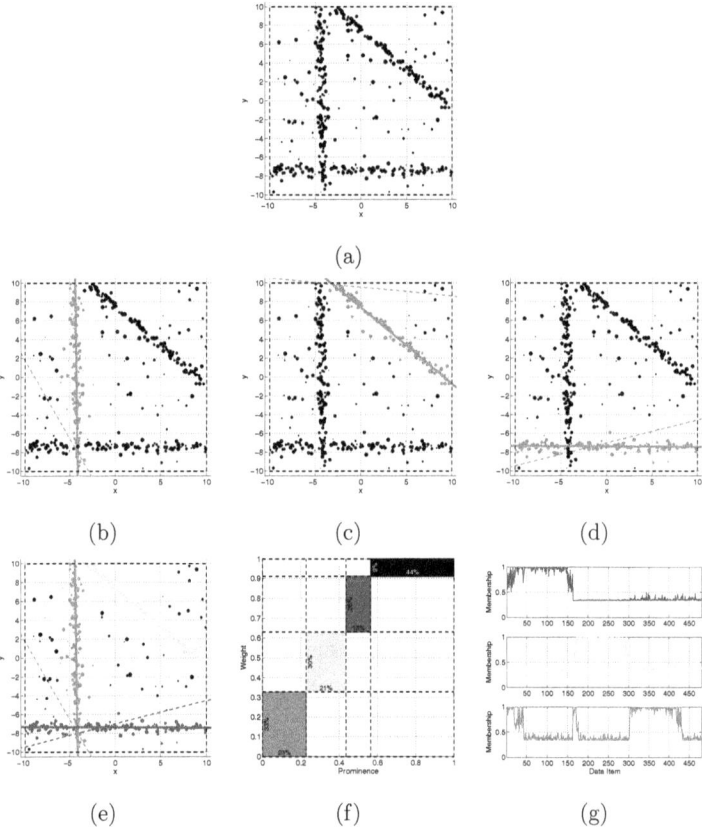

Figure 4.1: Different visualizations corresponding to a sample 2dl problem instance. (a) Input set of data items. (b), (c), (d) Members of $\bar{\Psi}$. (e) Members of $\bar{\Psi}$ super-positioned into the same axis system. (f) Prominence/Weight representation. (g) Membership of reordered data items to the members of $\bar{\Psi}$.

Table 4.1: Properties of the problem classes utilized in this paper.

Problem Class	x_n	ψ_c	Purpose	$\phi(x_n, \psi_c)$
2dc	$x_n \in \mathbb{R}^2$	$\psi_c = [m_c, \rho_c]$ $m_c \in \mathbb{R}^2$ $\rho_c > 0$	Finding Circles	$\left[\|x_n - m_c\|^2 - \rho_c^2\right]^2$
2de	$x_n \in \mathbb{R}^2$	$\psi_c \in \mathbb{R}^2$	Euclidean Clustering	$\|x_n - \psi_c\|^2$
2dl	$x_n \in \mathbb{R}^2$	$\psi_c = [m_c, v_c]$ $m_c \in \mathbb{R}^2$ $v_c \in \mathbb{R}^2, \|v_c\| = 1$	Finding Lines	$\|x_n - m_c - v_c^T(x_n - m_c)v_c\|^2$
3dpp	$x_n \in \mathbb{R}^3$	$\psi_c \in \mathbb{R}^3$	Finding Planes	$\dfrac{1}{\|\psi_c\|^2}\left(\psi_c^T x_n - \|\psi_c\|^2\right)^2$
ics	$x_n \in \mathbb{R}^3$	$\psi_c = [m_c, v_c]$ $m_c \in \mathbb{R}^3$ $v_c \in \mathbb{R}^3, \|v_c\| = 1$	Color Image Segmentation	$\|x_n - m_c - v_c^T(x_n - m_c)v_c\|^2$
ighe	$x_n \in \mathbb{R}$	$\psi_c \in \mathbb{R}$	Grayscale Image Segmentation	$(x_n - \psi_c)^2$

Figure 4.1(f) provides a visual representation of the prominence and weight values corresponding to the members of $\bar{\boldsymbol{\Psi}}$, i.e. $\tau_{\psi_{\bar{c}}}$ and $\omega_{\psi_{\bar{c}}}$, respectively. In this visualization, each box corresponds to one member of $\bar{\boldsymbol{\Psi}}$, except for the top right box which denotes "others". As such, the first 3 boxes, each have the width of $\tau_{\psi_{\bar{c}}}$ and height of $\omega_{\psi_{\bar{c}}}$, for $\bar{c} = 1, \cdots \bar{C}$. The top right box, on the other hand, has the width of $1 \quad \sum_{\bar{c}=1}^{\bar{C}} \tau_{\psi_{\bar{c}}}$ and height of $1 - \sum_{\bar{c}=1}^{\bar{C}} \omega_{\psi_{\bar{c}}}$. As such, the height of this box denotes the ratio of the members of $\check{\boldsymbol{\Psi}}$ which did not enter the aggregation process because their weight was less than θ_ω. The width of this box identifies the ratio of the members of $\check{\boldsymbol{\Psi}}$ which did aggregate into clusters, but were not prominent enough to be considered members of $\bar{\boldsymbol{\Psi}}$. In other words, the width of the top right box denotes all *locally optimal* clusters which did not occur frequently enough.

Finally, Figure 4.1(g) shows membership values corresponding to the input set of data items, wherein the data items are reordered using the following *index*,

$$i_n = \sum_{\bar{c}=1}^{\bar{C}} 2^{\bar{c}-1}\left[p_{n\bar{c}} > \frac{1}{2}\right], \tag{4.1}$$

We note that this membership output has similarities to the profile graphs generated by reVAT. Nevertheless, these graphs are generated by the developed method *after* the clustering process is

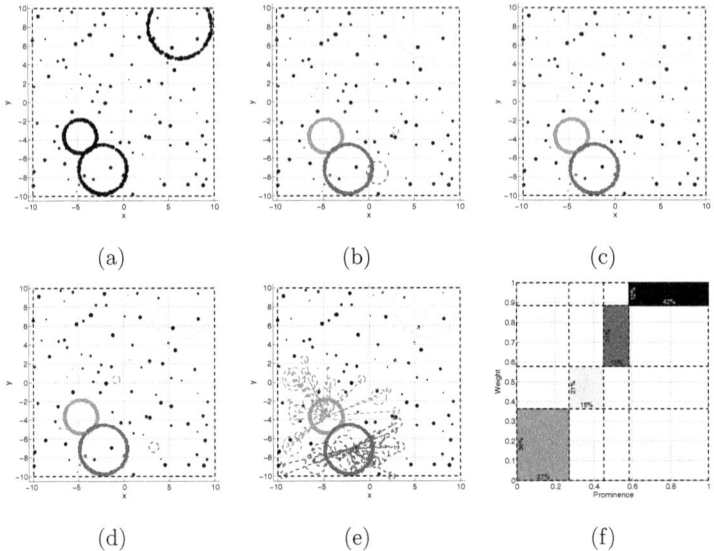

(a)   (b)   (c)

(d)   (e)   (f)

Figure 4.2: Repeatability study on a sample 2dc problem instance. (a) Input set of data items. (b)-(d) Three of the 25 repetitions of the proposed method. (e) Super-position of the results generated by the proposed method. (f) Prominence/Weight visualization corresponding to the solution carried in (b).

carried out. reVAT, on the other hand, performs thresholding on the relational Dissimilarity Matrix in order to generate the membership graphs. Hence, not only the reVAT approach is limited to relational problem classes, but also, and more importantly, the membership graphs in the present work are mere byproducts of the process employed by the developed algorithm and are by no means instrumental in the operation.

As stated, the proposed method utilizes several iterations of a local search mechanism in order to find prominent clusters relevant to an input set of data items. Hence, there is no implied guarantee of repeatability in the results of the proposed method for a unique set of input data items. Nevertheless, the clusters generated by the proposed method are sorted based on their corresponding value of prominence. Additionally, it is meaningful and even desirable to expect some level of repeatability in the outputs of the proposed method. Here, we describe an experiment which assesses the repeatability of the results of the proposed method.

Figure 4.2 carries the results of 25 independent executions of the proposed method for a sample

2dc problem instance. This problem class involves finding circular patterns in a weighted set of points in $\mathbb{R}^2$. In this figure, Figure 4.2(a) shows the input set of data items and Figures 4.2(b)-(d) show the results of three of the executions. Figure 4.2(e) carries the supper-position of the output generated by the 25 independent executions.

We first note that the proposed algorithm does in fact converge to 3 clusters in every execution. This is a direct result of the fact that this input set of data items contains 3 artificially created clusters. Under more realistic circumstances, we expect the number of clusters generated by the proposed method to be more varied between the different executions. Nevertheless, the rest of this experiment is applicable to a more general case as well.

Additionally, the ordering of the clusters generated by the proposed method is not necessarily repeatable. In fact, as stated above, the clusters are ordered by their corresponding values of prominence, i.e. clusters which appear more frequently are added higher in the list of clusters. Nevertheless, it is quite possible, and in fact it does occur frequently, that different executions of the developed algorithm produce the clusters in slightly different orders. Hence, in this experiment, we reorder the results generated at every execution so that maximum overlap between classification results is achieved. In other words, given two sets of clustering results $\{\tilde{\mathbf{X}}_{11}, \cdots \tilde{\mathbf{X}}_{1\bar{C}}\}$ and $\{\tilde{\mathbf{X}}_{21}, \cdots \tilde{\mathbf{X}}_{2\bar{C}}\}$, we reorder the second set so that $|\tilde{\mathbf{X}}_{1\bar{c}} \cap \tilde{\mathbf{X}}_{2\bar{c}}|$ is maximum for every $\bar{c}$. We follow this process using a greedy algorithm which iterates for $\bar{c} = 1 \cdots \bar{C}$ and finds the best match at each step. Note that the results of the proposed clustering algorithm are indifferent to reordering and, that, this post-processing step is carried out merely in order to allow for more meaningful comparison.

We define the dissimilarity between the results of two independent executions of the proposed method on a unique set of data items as the relative weight of the data items which are classified differently in the two executions. In other words, for the two sets of clusters $\bar{\mathbf{\Psi}}_1$ and $\bar{\mathbf{\Psi}}_2$, we define,

$$\delta_{\bar{\mathbf{\Psi}}_1, \bar{\mathbf{\Psi}}_2} = \frac{1}{\Omega} \sum_{n=1}^{N} \omega_n \left[c_{1n} \neq c_{2n} \right]. \tag{4.2}$$

As such, $\delta_{\bar{\mathbf{\Psi}}_1, \bar{\mathbf{\Psi}}_2}$ is zero iff $\bar{\mathbf{\Psi}}_1 = \bar{\mathbf{\Psi}}_2$, and is one, if no x_n is classified to the same cluster by the two solutions $\bar{\mathbf{\Psi}}_1$ and $\bar{\mathbf{\Psi}}_2$. Calculation of cluster dissimilarities for the 25 sets of clusters generated in the experiment carried in Figure 4.2, we observe that $\delta_{\bar{\mathbf{\Psi}}_1, \bar{\mathbf{\Psi}}_2} \leq 0.002315$ for every pair of $\bar{\mathbf{\Psi}}_1$ and $\bar{\mathbf{\Psi}}_2$. Given that the input set of data items in this problem instance includes 762 data items, this figure translates into a maximum of 1.76 discrepancies between the classifications carried out for the individual data items by the clustering solutions generated by the 25 independent executions of the proposed method.

Figure 4.2(e) shows the super-position of the 25 independent results generated by the developed algorithm. Here, as expected from the analysis carried above, we notice that the resulting clusters are coincident. While this confirms the numerical evaluation, it is informative to examine the dashed circles in Figure 4.2(e). Each one of these circles, denotes the seed cluster which resulted in one of the clusters which remained in the output set of clusters for one execution of the proposed method. Each one of these circles is connected to the corresponding output cluster through a dashed line. Hence, the star-shaped lines formed around each cluster in Figure 4.2(e) denote the set of clusters which converge to that cluster. Conceptually speaking, each one of these star-shaped geometries highlights the "basin" of a cluster, i.e. the "area" in which seed clusters tend to converge to the same "point". Given that the seeding mechanism utilized in this paper sweeps the cluster space using i.i.d. processes, the "area" covered by each star is also related to the prominence of the correspsonding cluster, as shown in Figure 4.2(f).

The proposed method in average requires $2,493 \pm 103$ milliseconds to converge for the 25 independent executions carried out in this experiment.

One key parameter which governs both the computational cost of the proposed algorithm and also the quality of the results generated by it, is the number of seed clusters processed by the proposed method, i.e. $\hat{C}$. In fact, analysis shows that the computational complexity of the proposed algorithm can be estimated as,

$$\hat{C} N \left[(5 + \chi_\phi + \chi_\Psi) i + 1 + 2\bar{C} \right]. \tag{4.3}$$

Here, i is the number of iterations that the proposed algorithm requires in average to converge, χ_ϕ is the cost of one time execution of the function $\phi(\cdot)$, and χ_Ψ is the cost of one time execution of the function $\Psi(\cdot)$ for one data item. Hence, defining $\chi = 5 + \chi_\phi + \chi_\Psi$, and assuming that $i\chi \gg 2\bar{C} + 1$, one can estimate the computational complexity of the proposed method as $\hat{C} N \chi i$. Here, χ and i are properties of the problem class and are independent of C and N. Hence, as also stated at the end of Section 3.6, the computational complexity of the proposed method is linear in terms of N; the number of data items in the problem instance. Also, note that the computational complexity of the proposed method scales linearly with $\bar{C}$; the reported number of clusters present in the set of data items. Additionally, (4.3) shows that cluster aggregation is in fact a less expensive operation that generating the locally optimal clusters.

While the cost of the proposed method scales linearly with $\hat{C}$, it is meaningful to expect that

Figure 4.3: Standard image *Lake*.

the quality of the results generated by it goes up as $\hat{C}$ increases. This *expectation* is based on the understanding that the proposed method "examines" $\hat{C}$ cluster seeds and aggregates them into a list of "unique" clusters. Hence, metaphorically speaking, a larger value of $\hat{C}$ translates into a "better" coverage of the search space. In other words, one would generate the "best" set of clusters if one can afford to let $\hat{C} \to \infty$.

Here, we carry the results of an experiment that investigates the impact of $\hat{C}$ on the quality of the results generated by the proposed method on an ighe problem instance and also examines the computational complexity of the proposed method as $\hat{C}$ increases.

Figure 4.4 shows the results of executing the proposed algorithm on the ighe problem instance which utilizes the standard image *Lake* shown in Figure 4.3 for different values of $\hat{C}$. Every item in this figure contains the histogram of the input image, at the top, and the square representation of the prominence/weight pairs corresponding to the clusters generated for the corresponding value of $\hat{C}$.

The results shown in Figure 4.4(a) correspond to $\hat{C} = 1$. In other words, in this case, $\hat{\Psi}$ contains one cluster, which ends up being *the* cluster generated by the proposed method as well. Hence, there is only one block in Figure 4.4(a)-bottom, the width of which, i.e. the prominence of the corresponding cluster is one. This cluster contains around 10% of the data items, hence the height of the only block in Figure 4.4(a)-bottom is 0.1. An increase of $\hat{C}$ to 2 does not change the situation, because, as seen in Figure 4.4(b), the two members of $\hat{\Psi}$ in fact converge to the same cluster. Hence, we observe one block with a prominence of one in Figure 4.4(b)-bottom as well. As seen in Figures 4.4(c)-(h), however, as $\hat{C}$ increases from 5 to 1,000, neither the output clusters nor their prominence and weight significantly differ for different values of $\hat{C}$. In fact, we observe in

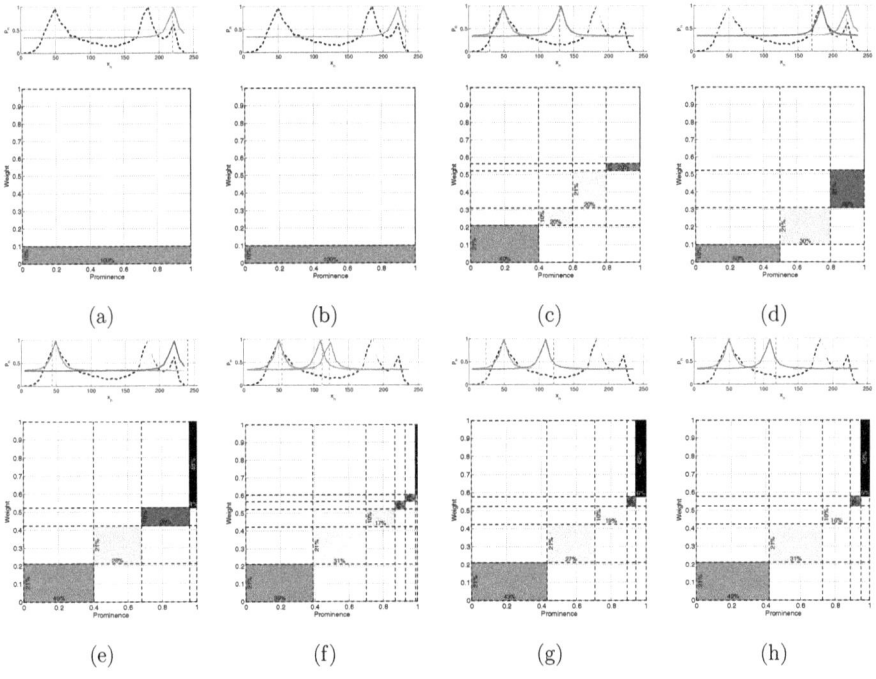

Figure 4.4: Output of the proposed method as $\hat{C}$ increases from 1 to $1{,}000$ for the ighe problem instance which utilizes the standard image *Lake* shown in Figure 4.3. (a) $\hat{C} = 1$. (b) $\hat{C} = 2$. (c) $\hat{C} = 5$. (d) $\hat{C} = 10$. (e) $\hat{C} = 25$. (f) $\hat{C} = 100$. (g) $\hat{C} = 500$. (h) $\hat{C} = 1{,}000$.

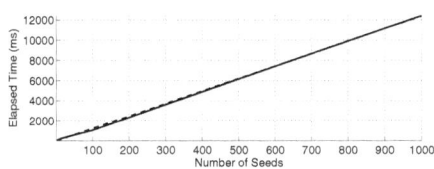

Figure 4.5: Time elapsed by the proposed algorithm for generating the results carried in Figure 4.4.

Figures 4.4(c)-(h) that always the three peaks of the histogram are discovered as three prominent clusters. Additionally, for some values of $\hat{C}$, one or two clusters of less prominence and weight are discovered at the middle of the histogram.

While an increase of $\hat{C}$ by a factor of 200 does not appear to change the results generated by the developed algorithm, the computational cost of the proposed algorithm is directly affected by the value of $\hat{C}$. As observed in Figure 4.5, the elapsed time of the proposed algorithm for different values of $\hat{C}$ grows linearly as $\hat{C}$ increases. This is in direct accordance with the analysis carried above.

Hence, we make the following two observations. First, increasing $\hat{C}$ does not deteriorate the quality of the results generated by the developed algorithm. Second, the time required by the proposed algorithm in order to converge inflates linearly as $\hat{C}$ increases. As such, the choice for $\hat{C}$ is a *budgetary* one. In other words, one has to allocate *enough* resources in order for the proposed algorithm to discover the clusters present in the input set of data items, whereas increasing $\hat{C}$ beyond this requirement merely results in the reinforcement of the decision to converge to the discovered clusters. Hence, one may incorporate budgeting strategies such as "stop when no new cluster has been discovered in a predetermined interval", among others, into the proposed algorithm.

It is important to note that the particular number of clusters present in any of the results shown in Figure 4.4 is in fact only indicative of the particular prominence threshold which is used in every case. In other words, the proposed method produces some number of clusters $\bar{C}$, for which we only know that $\bar{C} \leq \check{C}$. This is in fact due to the mechanics of the production and aggregation of $\bar{\boldsymbol{\Psi}}$. Hence, as discussed above, as one increases $\hat{C}$ towards infinity, one must expect, theoretically at least, to receive an infinite number of clusters, each of which is to some level prominent. This phenomenon can be recognized in Figure 4.6(b) wherein θ_τ is allowed to drop to 0.001, and, therefore, for any cluster to pass (here $\hat{C} = 255$, therefore any single cluster is a *prominent* aggregate cluster). Note that the fact that there are 10 clusters in $\bar{\boldsymbol{\Psi}}$, for the case shown

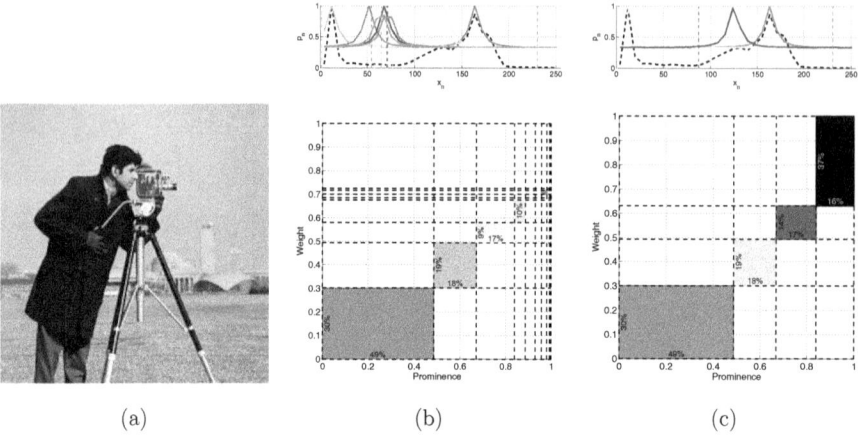

(a)   (b)   (c)

Figure 4.6: Results of an experiment on the minimum cluster prominence threshold. (a) Standard image *Cameraman*. (b) $\theta_\tau = 0.001$. (b) $\theta_\tau = 0.1$.

in Figure 4.6(b), is merely because $\bar{C}$ is capped at 10, due to visualization requirements. In the following paragraph, we review the experiment carried in Figure 4.6.

Figure 4.6(a) shows the standard image *Cameraman*. This image is fed to the proposed algorithm, within the context of an ighe problem instance, and after $2,719$ milliseconds of processing, the 10 clusters carried in Figure 4.6(b) are generated. Note the overlaps between many of these clusters and also the sequence of clusters with negligible prominence and weight factors present in the prominence/weight visualization.

When we apply a minimum prominence threshold of 10%, i.e. $\theta_\tau = 0.1$, cluster count drops to 3, as shown in Figure 4.6(c). We suggest that this set of clusters constitutes a more *appropriate* solution to this problem instance, due to the fact that the clusters contained in this solution are more prominent. The effects of this enforcement are visible, both in the histogram domain as well as the prominence/weight visualization.

Finally, we present problem instances corresponding to the problem classes listed in Table 4.1. Figures 4.7 and 4.8 carry these experiments. In each case, the top item shows the input set of data items and the middle item shows the results generated by the proposed method. The bottom item in each case shows the corresponding prominence/weight visualization. In case of Figure 4.8 an additional row visualizes the clusters in their own domain. Note that in the six experiments exhibited in Figures 4.7 and 4.8, the value of θ_τ is set so that three clusters are generated by the

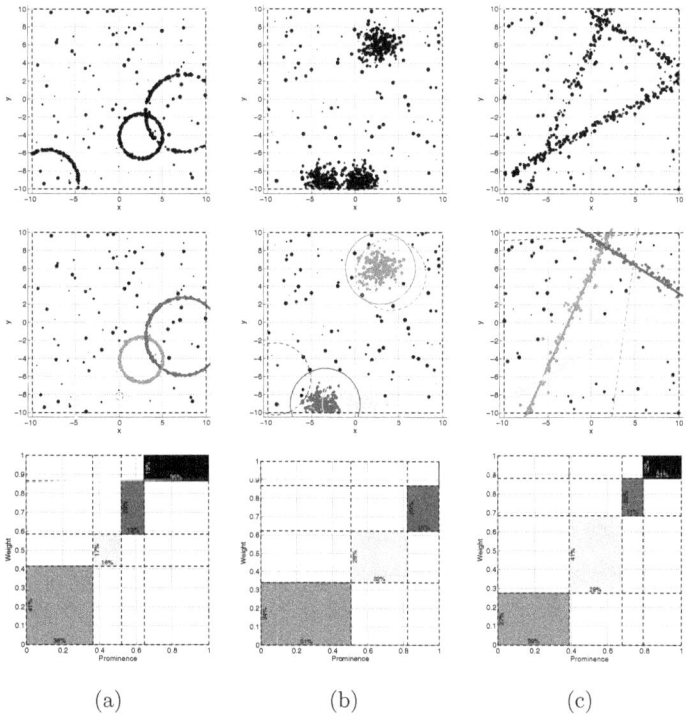

Figure 4.7: Results of the proposed method for problem instances corresponding to the problem classes listed in Table 4.1. (a) 2dc. (b) 2de. (c) 2dl.

proposed method. This is *merely* in order to produce more pleasing visual outputs. In fact, as discussed frequently in the preceding parts of this paper, the choice of how far into the list of clusters generated by the proposed method we go for the purpose of visualization and reporting, has *no* impact on the particular clusters present in the solution generated by the proposed method.

Figure 4.7(a) shows a 2dc problem instance. We note that the circle at the middle of the working domain is the most prominent cluster. This is anticipated because of a phenomenon similar to the one exhibited in Figure 4.2(e). As such, the cluster with a higher probability of other clusters landing in its vicinity is more likely to appear more prominently in $\check{\Psi}$. This problem instance contains 631 data items and it takes the proposed method $4,891$ milliseconds to converge to the 3 clusters shown in Figure 4.7(a).

Similarly, Figure 4.7(b) shows a 2de problem instance, wherein there exists a cluster which is farther than the two others from the edges of the working domain. We observe, again, that this cluster appears more prominently in the output set of clusters as well. This problem instance contains 723 data items and it takes the proposed method $3,344$ milliseconds to converge to the 3 clusters shown in Figure 4.7(b).

In the case of Figure 4.7(c), there are 432 data items in this problem instance and the proposed method takes $2,750$ milliseconds to converge to the 3 clusters shown in Figure 4.7(c).

The three problem instances carried in Figure 4.7 correspond to artificially generated sets of data items. In these cases, there in fact exists a value of C which is well applicable to each problem instance, i.e. the value of C which is utilized for generating $\mathbf{X}$ in the first place. It is important to emphasize that the proposed algorithm is not aware of this value, but, nevertheless, there does exist a value of C which is very well applicable to any problem instance corresponding to these and other problem classes which utilize artificial data. The problem instances shown in Figure 4.8, however, belong to problem classes which process real data captured from real physical processes.

Figure 4.8(a) shows a 3dpp problem instance. The input set of data items in this problem class contains 3D points captured by a *Kinect 2* sensor. The depth-maps used in this experiment are captured at the resolution of 424×512 pixels. Here, intrinsic parameters of the camera are acquired through the Kinect SDK. The scene captured in this data corresponds to the corner of a room which contain three human bodies. We note that the proposed method finds 3 clusters here, i.e. the 3 walls of the room. The weights and the prominence values corresponding to these 3 clusters are almost identical. Also note that about half of the seed clusters converge to clusters which are not prominent enough to be considered in the set of clusters reported by the proposed

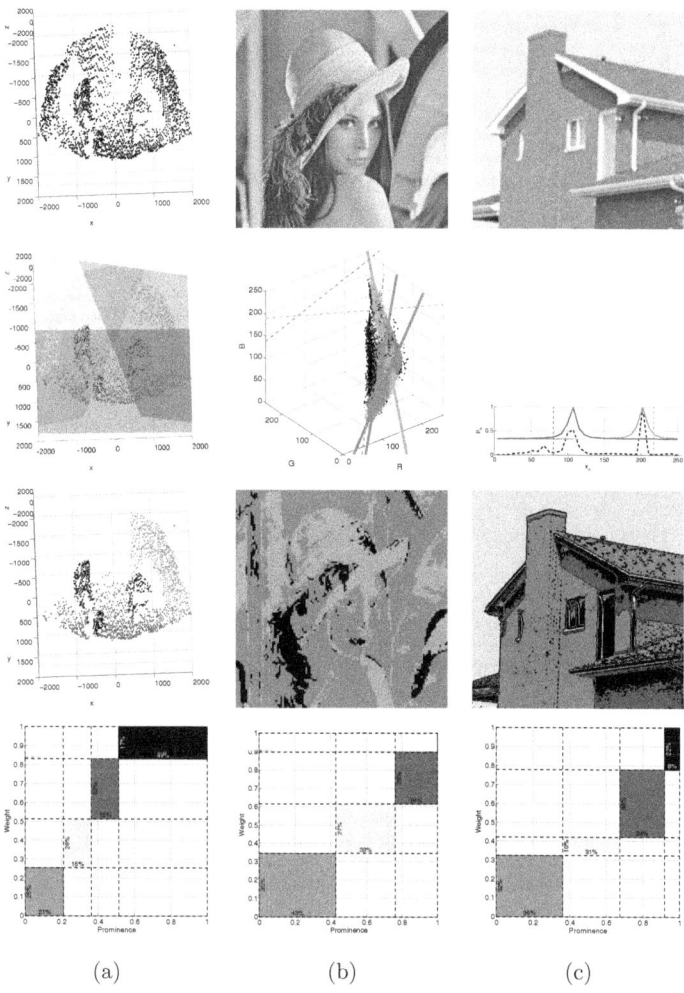

Figure 4.8: Results of the proposed method for problem instances corresponding to the problem classes listed in Table 4.1. (a) 3dpp. Kinect data corresponding to a corner of room with human bodies present in the scene. Data courtesy of *Epson Edge, Epson Canada Limited*. (b) ics. Standard image *Lena*. (c) ighe. Standard image *House*.

method. Additionally, about one sixth of the data points are not classified into the 3 clusters discovered by the proposed method. The set of input data items utilized in this problem instance contains $2,708$ points and the proposed method requires $22,906$ milliseconds to produce the output carried in Figure 4.8(a).

The results shown in Figures 4.8(b) and (c) correspond to ics and ighe problem instances. These two problem classes involve the segmentation of color and grayscale images, respectively. Note that neither of these problem classes utilize the spatial context of the data items. The reader is referred to provisions related to the spatial context in [98].

Figure 4.8(b) shows the classification of the color vectors in the standard image *Lena* using a linear model of homogeneity [99] (also see [100]). We note that, in this case, almost every cluster seed converges to one of the clusters present in $\bar{\boldsymbol{\Psi}}$. The weight visualization carried in Figure 4.8(b), however, shows that about 10% of the data items are not classified into any of the members of $\bar{\boldsymbol{\Psi}}$. The set of data items utilized in this problem instance contains $16,384$ vectors in $\mathbb{R}^3$ and the proposed method requires $32,641$ milliseconds to produce the 3 clusters carried in Figure 4.8(b).

Finally, Figure 4.8(c) shows the results generated by the proposed algorithm on an ighe problem instance. Here, the standard image *House* is used. The histogram of the grayscale values corresponding to this image shows 3 distinct peaks. The proposed method, too, converges to these 3 clusters. We note that these clusters are all equally prominent, while the middle clusters carries almost a third of the weight of the two others. This is expected due to the lower area underneath the section of the histogram which corresponds to this cluster, compared to the two others. This set of data items contains 32 bins of the histogram and the proposed method requires $2,250$ milliseconds to converge for it.

Neither FCM, nor PCM, or any of their many variants, hybrids, and alternatives, are compared to the method developed in this research. This is neither a coincidence, nor the result of an underestimation of the magnificent value of these and other important contributions to the field.

We refrained from comparing the developed method with what is available in the literature, because this work does not, either implicitly or explicitly, claim to propose an alternative to any of those algorithms. In fact, there is plenty of evidence in the literature that a practitioner of the field can find and tune-up one of the many algorithms proposed in the past few decades and "make it work" for their immediate need. In fact, there exist implementations of those algorithms in major programming languages, including C++, Matlab/Octave, and Python, which are practically useful for many real-world problems. Henceforth, we decidedly avoid the, to our understanding, *fruitless* comparison of the present work with that section of the literature.

The focus of this work is on the existence of C, and more importantly, on the assumption of the validity of the assumption that there is a C. As such, FCM, and a great many of its successors, not only assume that there exists a C which is "the C", but that it can be known either *a priori* or *a posteriori*. In this context, this work is different from both paradigms in the sense that it asserts that *there is no C*. In other words, there is no number of clusters which is *correct* for a problem instance which involves data collected in the physical world. As such, the problem is not that C is not known, which is what FCM is afflicted with, and not that estimating C is hard, which is what VAT tries to provide a solution for, but that there is no C and that the model must not include any *hard* notion of "the number of clusters".

Hence, as discussed before, we dethrone C from the supreme role of moderating the number of clusters present in the system and show that it is really the user who must be able to decide when the stream of clusters must stop. This stream must contain clusters at descending order of prominence, each of which is nevertheless to some level relevant to the data. Thus, we assert that the fact that a particular user decides to discard the outputs of the algorithm at some point is in fact her call and not something that the clustering algorithm must and should be aware and in charge of.

The proposed algorithm, as such, is a cluster discovery machinery which utilizes the allotted budget in order to aggregate clusters which are locally optimal for the input set of data items and returns them in descending order of prominence. the user decides for how long this machinery must work and what portion of its outputs are usable. We are not aware of a counterpart for this process, and, hence, we are incapable of comparing the developed algorithm with any one of the previous contributions in the field. It is needless to say that those contributions are the important pillars which have enabled the development of the present work.

Chapter 5

Conclusions

In this work we address a key assumption in the field of fuzzy clustering and propose that, contrary to the history and convention of the field, it is not meaningful to discuss "the number of clusters present in a set of data items". This is an invaluable suggestion, because, the first major historical steps in the field, including the pioneering work on FCM, make the assumption that it is known through some other process that there are C clusters in the input set of data items $\mathbf{X}$. FCM, and its many predecessors and incarnations and hybrids, then search for C clusters in $\mathbf{X}$. In addition to other difficulties, the community soon recognized that generating an appropriate C for an arbitrary set $\mathbf{X}$ is not a trivial task. The challenges of estimating C are in fact more severe and the impact of an improper C is more critical when one deploys unsupervised data clustering on non-relational problem classes. We argue that the mere notion that a deterministic value of C exists for an arbitrary problem instance is erroneous. In other words, the problem is *not* that one does not know what C is for an arbitrary problem instance. In fact, we argue that the problem is *not* the extended problem that the practitioner of the field does not have access to a solid estimation mechanism for the value of C within the context of a generic notion of homogeneity (to which VAT is not applicable) either. To our understanding, the struggle for "finding/using the proper C" arises from the fact that there is no C. In other words, when problem classes involve data captured from actual physical processes, there is no value of C which defines the number of clusters present in $\mathbf{X}$. Hence, we assert, the clustering algorithm must be indifferent to any notion of "the number of clusters" as a prerequisite for the clustering effort. We demonstrate that this novel paradigm is possible when a robustified single-cluster clustering algorithm is executed in order to generate a large set of locally optimal cluster representations which are then aggregated into a few clusters indexed through their prominence values. We show that the entirety of this process can be executed agnostic of the particular data item and cluster models relevant to any problem class. As such, in this paper, we propose a method which is capable of discovering clusters in a given set of data items to which an arbitrary notion of homogeneity is applicable.

Acknowledgment

We wish to thank the management of *Fio Corporation* for their support. We thank the management of *Epson Edge, Epson Canada Limited*, for allowing us to use the Kinect data utilized in some of the experiments carried in this paper. The author wishes to thank *Professor James C. Bezdek* for his mentorship. The majority of this research was carried out in *Istanbul Cafe & Espresso Bar* and *Rooster Coffee House (King East)*, both located in Toronto. The author wishes to thank *Elham Nemati* for assisting us with finding key pieces of the literature for this work and for proofreading this manuscript.

Bibliography

[1] Twitter's IPO prospectus, Filed with the Securities and Exchange Commission (3 October 2013).
URL `https://goo.gl/lY6JXG`

[2] S. Pigg, House sales in GTA close to setting record, despite listings shortage, Toronto Star (7 January 2015).
URL `https://goo.gl/IKdBMo`

[3] Photos, photos everywhere, The New York Times (29 July 2015).
URL `https://goo.gl/huZP7O`

[4] J. B. MacQueen, Some methods for classification and analysis of multivariate observations, in: Proceedings of 5-th Berkeley Symposium on Mathematical Statistics and Probability, Berkeley, 1967, pp. 281–297.

[5] R. Duda, P. Hart, Pattern Classification and Scene Analysis, Wiley, New York, 1973.

[6] R. Gray, Y. Linde, Vector quantizers and predictive quantizers for Gauss-Markov sources, IEEE Transactions on Communications 30 (2) (1982) 381–389.

[7] G. H. Ball, D. J. Hall, A clustering technique for summarizing multivariate data, Behavioral Science 12 (2) (1967) 153–155.

[8] H. Cheng, J.-R. Chen, J. Li, Threshold selection based on fuzzy c-partition entropy approach, Pattern Recognition 31 (7) (1998) 857–870.

[9] J. C. Bezdek, Numerical taxonomy with fuzzy sets, Journal of Mathematical Biology 1 (1) (1974) 57–71.

[10] J. C. Bezdek, J. Dunn, Optimal fuzzy partitions: A heuristic for estimating the parameters in a mixture of normal distributions, IEEE Transactions on Computers C-24 (8) (1975) 835–838.

[11] J. C. Bezdek, A physical interpretation of fuzzy ISODATA, IEEE Transactions on Systems, Man and Cybernetics SMC-6 (5) (1976) 387–389.

[12] J. C. Bezdek, J. M. Keller, R. Krishnapuram, N. R. Pal, Fuzzy Models and Algorithms for Pattern Recognition and Image Processing, Kluwer Academic Publishers, Boston, 1999.

[13] S. Miyamoto, Fuzzy clustering - Basic ideas and overview, in: J. Kacprzyk, W. Pedrycz (Eds.), Springer Handbook of Computational Intelligence, Springer Berlin Heidelberg, 2015, pp. 239–248.

[14] E. H. Ruspini, A new approach to clustering, Information & Control 15 (1) (1969) 22–32.

[15] J. C. Dunn, A fuzzy relative of the ISODATA process and its use in detecting compact well-separated clusters, Journal of Cybernetics 3 (3) (1973) 32–57.

[16] J. C. Bezdek, Pattern Recognition with Fuzzy Objective Function Algorithms, Plenum Press, New York, 1981.

[17] J. M. Leski, Generalized weighted conditional fuzzy clustering, IEEE Transactions on Fuzzy Systems 11 (6) (2003) 709–715.

[18] J. Yu, Q. Cheng, H. Huang, Analysis of the weighting exponent in the FCM, IEEE Transactions on Systems, Man, and Cybernetics, Part B: Cybernetics 34 (1) (2004) 634–639.

[19] M. Trivedi, J. C. Bezdek, Low-level segmentation of aerial images with fuzzy clustering, IEEE Transactions on Systems, Man, and Cybernetics 16 (4) (1986) 589–598.

[20] N. R. Pal, J. C. Bezdek, On cluster validity for the fuzzy C-means model, IEEE Transactions on Fuzzy Systems 3 (3) (1995) 370–379.

[21] H. Frigui, R. Krishnapuram, A robust algorithm for automatic extraction of an unknown number of clusters from noisy data, Pattern Recognition Letters 17 (12) (1996) 1223–1232.

[22] F. Klawonn, R. Kruse, H. Timm, Fuzzy shell cluster analysis, in: G. della Riccia, H. Lenz, R. Kruse (Eds.), Learning, networks and statistics, Springer, 1997, pp. 105–120.

[23] C. Borgelt, Objective functions for fuzzy clustering, in: C. Moewes, A. Nurnberger (Eds.), Computational Intelligence in Intelligent Data Analysis, Vol. 445 of Studies in Computational Intelligence, Springer Berlin Heidelberg, 2013, pp. 3–16.

[24] F. Klawonn, F. Hoppner, What is fuzzy about fuzzy clustering? Understanding and improving the concept of the fuzzifier, in: M. R. Berthold, H.-J. Lenz, E. Bradley, R. Kruse, C. Borgelt (Eds.), Advances in Intelligent Data Analysis V, Vol. 2810 of Lecture Notes in Computer Science, Springer Berlin Heidelberg, 2003, pp. 254–264.

[25] I. H. Suh, J.-H. Kim, F. Chung-Hoon Rhee, Convex-set-based fuzzy clustering, IEEE Transactions on Fuzzy Systems 7 (3) (1999) 271–285.

[26] R. Kruse, C. Doring, M.-J. Lesot, Fundamentals of fuzzy clustering, in: J. V. de Oliveira, W. Pedrycz (Eds.), Advances in Fuzzy Clustering and its Applications, Wiley, England, 2007, pp. 3–29.

[27] R. Yager, D. Filev, Approximate clustering via the mountain method, IEEE Transactions on Systems, Man and Cybernetics 24 (8) (1994) 1279–1284.

[28] G. Beni, X. Liu, A least biased fuzzy clustering method, IEEE Transactions on Pattern Analysis and Machine Intelligence 16 (9) (1994) 954–960.

[29] K. Rose, E. Gurewitz, G. Fox, A deterministic annealing approach to clustering, Pattern Recognition Letters 11 (9) (1990) 589–594.

[30] K. Rose, E. Gurewitz, G. Fox, Constrained clustering as an optimization method, IEEE Transactions on Pattern Analysis and Machine Intelligence 15 (8) (1993) 785–794.

[31] J. M. Leski, Fuzzy c-varieties/elliptotypes clustering in reproducing kernel Hilbert space, Fuzzy Sets and Systems 141 (2) (2004) 259–280.

[32] C. Borgelt, C. Braune, M.-J. Lesot, R. Kruse, Handling noise and outliers in fuzzy clustering, in: D. E. Tamir, N. D. Rishe, A. Kandel (Eds.), Fifty Years of Fuzzy Logic and its Applications, Vol. 326 of Studies in Fuzziness and Soft Computing, Springer International Publishing, 2015, pp. 315–335.

[33] T. Kwok, K. Smith, S. Lozano, D. Taniar, Parallel fuzzy c-means clustering for large data sets, in: B. Monien, R. Feldmann (Eds.), Euro-Par 2002 Parallel Processing, Vol. 2400 of Lecture Notes in Computer Science, Springer Berlin Heidelberg, 2002, pp. 365–374.

[34] T. M. Nguyen, Q. Wu, Dynamic fuzzy clustering and its application in motion segmentation, IEEE Transactions on Fuzzy Systems 21 (6) (2013) 1019–1031.

[35] J. Zhou, C. P. Chen, L. Chen, H.-X. Li, A collaborative fuzzy clustering algorithm in distributed network environments, IEEE Transactions on Fuzzy Systems 22 (6) (2014) 1443–1456.

[36] F. Masulli, S. Rovetta, Soft transition from probabilistic to possibilistic fuzzy clustering, IEEE Transactions on Fuzzy Systems 14 (4) (2006) 516–527.

[37] R. Krishnapuram, J. M. Keller, The possibilistic C-means algorithm: insights and recommendations, IEEE Transactions on Fuzzy Systems 4 (3) (1996) 385–393.

[38] S. Miyamoto, D. Suizu, Fuzzy c-means clustering using kernel functions in support vector machines, Journal of Advanced Computational Intelligence and Intelligent Informatics 7 (1) (2003) 25–30.

[39] D.-M. Tsai, C.-C. Lin, Fuzzy C-means based clustering for linearly and nonlinearly separable data, Pattern Recognition 44 (8) (2011) 1750–1760.

[40] K.-L. Wu, M.-S. Yang, Alternative C-means clustering algorithms, Pattern Recognition 35 (10) (2002) 2267–2278.

[41] L. Kaufman, P. J. Rousseeuw, Finding Groups in Data: an Introduction to Cluster Analysis, John Wiley & Sons Inc, New York, 1990.

[42] R. J. Hathaway, J. W. Davenport, J. C. Bezdek, Relational duals of the C-means clustering algorithms, Pattern Recognition 22 (2) (1989) 205–212.

[43] R. J. Hathaway, J. C. Bezdek, NERF C-means: Non-Euclidean relational fuzzy clustering, Pattern Recognition 27 (3) (1994) 429–437.

[44] S. Nascimento, B. Mirkin, F. Moura-Pires, Multiple prototype model for fuzzy clustering, in: D. J. Hand, J. N. Kok, M. R. Berthold (Eds.), Advances in Intelligent Data Analysis, Vol. 1642 of Lecture Notes in Computer Science, Springer Berlin Heidelberg, 1999, pp. 269–279.

[45] L. Fu, E. Medico, FLAME, A novel fuzzy clustering method for the analysis of DNA microarray data, BMC Bioinformatics 8 (3).

[46] T. Hastie, R. Tibshirani, J. Friedman, The Elements of Statistical Learning, Springer, New York, 2009.

[47] D. E. Gustafson, W. C. Kessel, Fuzzy clustering with a fuzzy covariance matrix, in: IEEE Conference on Decision and Control including the 17th Symposium on Adaptive Processes, Vol. 17, San Diego, CA, 1979, pp. 761–766.

[48] I. Gath, A. Geva, Unsupervised optimal fuzzy clustering, IEEE Transaction on Pattern Analysis Machine Intelligence 11 (7) (1989) 773–781.

[49] R. J. Hathaway, J. C. Bezdek, Switching regression models and fuzzy clustering, IEEE Transactions on Fuzzy Systems 1 (3) (1993) 195–204.

[50] R. Babuska, P. van der Veen, U. Kaymak, Improved covariance estimation for Gustafson-Kessel clustering, in: Proceedings of the 2002 IEEE International Conference on Fuzzy Systems (FUZZ-IEEE 2002), Vol. 2, 2002, pp. 1081–1085.

[51] H. Frigui, R. Krishnapuram, A comparison of fuzzy shell-clustering methods for the detection of ellipses, IEEE Transactions on Fuzzy Systems 4 (2) (1996) 193–199.

[52] R. N. Dave, R. Krishnapuram, Robust clustering methods: A unified view, IEEE Transactions on Fuzzy Systems 5 (2) (1997) 270–293.

[53] K. K. Chintalapudi, M. Kam, The credibilistic fuzzy C-means clustering algorithm, in: IEEE International Conference on Systems, Man, and Cybernetics (SMC 1998), Vol. 2, 1998, pp. 2034–2039.

[54] N. R. Pal, K. Pal, J. M. Keller, J. C. Bezdek, A possibilistic fuzzy c-means clustering algorithm, IEEE Transactions on Fuzzy Systems 13 (4) (2005) 517–530.

[55] J. Leski, Towards a robust fuzzy clustering, Fuzzy Sets and Systems 137 (2) (2003) 215–233.

[56] P. D'Urso, L. D. Giovanni, Robust clustering of imprecise data, Chemometrics and Intelligent Laboratory Systems 136 (2014) 58–80.

[57] J. J. D. Gruijter, A. B. McBratney, A modified fuzzy K-means method for predictive classification, in: H. H. Bock (Ed.), Classification and Related Methods of Data Analysis, Elsevier, Amsterdam, The Netherlands, 1988, pp. 97–104.

[58] R. N. Dave, Characterization and detection of noise in clustering, Pattern Recognition Letters 12 (11) (1991) 657–664.

[59] Y. Ohashi, Fuzzy clustering and robust estimation, Presented at the 9th SAS Users Group International (SUGI) Meeting at Hollywood Beach, Florida. (1984).

[60] R. N. Dave, Robust fuzzy clustering algorithms, in: Second IEEE International Conference on Fuzzy Systems, Vol. 2, 1993, pp. 1281–1286.

[61] R. Krishnapuram, J. M. Keller, A possibilistic approach to clustering, IEEE Transactions on Fuzzy Systems 1 (2) (1993) 98–110.

[62] M. Barni, V. Cappellini, A. Mecocci, Comments on "A possibilistic approach to clustering", IEEE Transactions on Fuzzy Systems 4 (3) (1996) 393–396.

[63] H. Timm, C. Borgelt, C. Doring, R. Kruse, An extension to possibilistic fuzzy cluster analysis, Fuzzy Sets and Systems 147 (1) (2004) 3–16.

[64] N. R. Pal, K. Pal, J. C. Bezdek, A mixed c-means clustering model, in: Proceedings of the Sixth IEEE International Conference on Fuzzy Systems, Vol. 1, 1997, pp. 11–21.

[65] N. R. Pal, K. Pal, J. M. Keller, J. C. Bezdek, A new hybrid C-means clustering model, in: Proceedings of the 2004 IEEE International Conference on Fuzzy Systems, Vol. 1, 2004, pp. 179–184.

[66] A. Abadpour, Rederivation of the fuzzypossibilistic clustering objective function through Bayesian inference, Fuzzy Sets and Systems 305 (2016) 29–53.

[67] A. Abadpour, A sequential bayesian alternative to the classical parallel fuzzy clustering model, Information Sciences 318 (2015) 28–47.

[68] R. Krishnapuram, H. Frigui, O. Nasraoui, Fuzzy and possibilistic shell clustering algorithms and their application to boundary detection and surface approximation - part I, IEEE Transaction on Fuzzy Systems 3 (1) (1995) 29–43.

[69] R. Krishnapuram, H. Frigui, O. Nasraoui, Fuzzy and possibilistic shell clustering algorithms and their application to boundary detection and surface approximation - Part II, IEEE Transaction on Fuzzy Systems 3 (1) (1995) 44–60.

[70] R. N. Dave, T. Fu, Robust shape detection using fuzzy clustering: Practical applications, Fuzzy Sets and Systems 65 (2-3) (1994) 161–185.

[71] J. M. Jolion, P. Meer, S. Bataouche, Robust clustering with applications in computer vision, IEEE Transactions on Pattern Analysis and Machine Intelligence 13 (8) (1991) 791–802.

[72] I. Sledge, J. C. Bezdek, T. C. Havens, J. M. Keller, Relational generalizations of cluster validity indices, IEEE Transactions on Fuzzy Systems 18 (4) (2010) 771–786.

[73] H. Frigui, R. Krishnapuram, Clustering by competitive agglomeration, Pattern Recognition 30 (7) (1997) 1109–1119.

[74] J. C. Dunn, Indices of partition fuzziness and the detection of clusters in large data sets, in: B. G. M.M. Gupta, G.N. Saridis (Ed.), Automata and Decision Process, North-Holland Publishers, Amsterdam, 1977, pp. 271–283.

[75] L. Hubert, P. Arabie, Comparing partitions, Journal of Classification 2 (1) (1985) 193–218.

[76] D. L. Davies, D. W. Bouldin, A cluster separation measure, IEEE Transactions on Pattern Analysis and Machine Intelligence PAMI-1 (2) (1979) 224–227.

[77] M. K. Pakhira, S. Bandyopadhyay, U. Maulik, Validity index for crisp and fuzzy clusters, Pattern Recognition 37 (3) (2004) 487–501.

[78] J. C. Bezdek, Q. W. Li, Y. Attikiouzel, M. Windham, A geometric approach to cluster validity for normal mixtures, Soft Computing 1 (4) (1997) 166–179.

[79] R. J. Hathaway, J. C. Bezdek, Visual cluster validity for prototype generator clustering models, Pattern Recognition Letters 24 (9–10) (2003) 1563–1569.

[80] J. C. Bezdek, R. J. Hathaway, VAT: A tool for visual assessment of (cluster) tendency, in: Proceedings of the 2002 International Joint Conference on Neural Networks (IJCNN 2002), Vol. 3, 2002, pp. 2225–2230.

[81] L. Freeman, Displaying hierarchical clusters, Connections 17 (2) (1994) 46–52.

[82] P. J. Rousseeuw, Silhouettes: A graphical aid to the interpretation and validation of cluster analysis, Journal of Computational and Applied Mathematics 20 (1987) 53–65.

[83] T. Tran-Luu, Mathematical concepts and novel heuristic methods for data clustering and visualization, Ph.D. thesis, University of Maryland, College Park, MD (1996).

[84] L. Wang, X. Geng, J. C. Bezdek, C. Leckie, R. Kotagiri, SpecVAT: Enhanced visual cluster analysis, in: Proceedings of 2008 Eighth IEEE International Conference on Data Mining, 2008, pp. 638–647.

[85] M. Belkin, P. Niyogi, Laplacian eigenmaps and spectral techniques for embedding and clustering, in: T. G. Dietterich, S. Becker, Z. Ghahramani (Eds.), Advances in Neural Information Processing Systems 14, 2001, pp. 585–591.

[86] L. Wang, U. T. Nguyen, J. C. Bezdek, C. A. Leckie, K. Ramamohanarao, iVAT and aVAT: Enhanced visual analysis for cluster tendency assessment, in: M. J. Zaki, J. X. Yu, B. Ravindran, V. Pudi (Eds.), Advances in Knowledge Discovery and Data Mining, Vol. 6118 of Lecture Notes in Computer Science, Springer Berlin Heidelberg, 2010, pp. 16–27.

[87] J. M. Huband, J. C. Bezdek, R. J. Hathaway, Revised visual assessment of (cluster) tendency (reVAT), in: Proceedings of IEEE Annual Meeting of the Fuzzy Information (NAFIPS 2004), Vol. 1, 2004, pp. 101–104.

[88] J. M. Huband, J. C. Bezdek, R. J. Hathaway, bigVAT: Visual assessment of cluster tendency for large data sets, Pattern Recognition 38 (11) (2005) 1875–1886.

[89] R. J. Hathaway, J. C. Bezdek, J. M. Huband, Scalable visual assessment of cluster tendency for large data sets, Pattern Recognition 39 (7) (2006) 1315–1324.

[90] T. C. Havens, J. C. Bezdek, M. Palaniswami, Scalable single linkage hierarchical clustering for big data, in: Proceedings of 2013 IEEE Eighth International Conference on Intelligent Sensors, Sensor Networks and Information Processing, 2013, pp. 396–401.

[91] V. Ganti, J. Gehrke, R. Ramakrishnan, Mining very large databases, Computer 32 (8) (1999) 38–45.

[92] B. Fischer, T. Zoller, J. M. Buhmann, Proceedings of the third international workshop on energy minimization methods in computer vision and pattern recognition (emmcvpr 2001),

Springer Berlin Heidelberg, Berlin, Heidelberg, 2001, Ch. Path Based Pairwise Data Clustering with Application to Texture Segmentation, pp. 235–250.

[93] D. Kumar, J. C. Bezdek, M. Palaniswami, S. Rajasegarar, C. Leckie, T. C. Havens, A hybrid approach to clustering in big data, IEEE Transactions on Cybernetics PP (99) (2015) 1–1. doi:10.1109/TCYB.2015.2477416.

[94] R. Sibson, SLINK: An optimally efficient algorithm for the single-link cluster method, The Computer Journal 16 (1) (1973) 30–34.

[95] P. W. Holland, R. E. Welsch, Robust regression using iteratively reweighted least squares, Communication Statistics - Theory and Methods A6 (9) (1977) 813–827.

[96] G. Wesolowski, The Weber problem: History and perspective, Location Science 1 (1993) 5–23.

[97] J. K. Lindsey, Comparison of probability distributions, Journal of the Royal Statistical Society. Series B (Methodological) 36 (1) (1974) 38–47.

[98] A. Abadpour, Incorporating spatial context into fuzzy-possibilistic clustering using Bayesian inference, Journal of Intelligent & Fuzzy Systems 30 (2) (2016) 895–919.

[99] G. J. Klinker, S. A. Shafer, T. Kanade, A physical approach to color image understanding, International Journal of Computer Vision 4 (1990) 7–38.

[100] A. Abadpour, S. Kasaei, Color PCA eigenimages and their application to compression and watermarking, IEE Image & Vision Computing 26 (7) (2008) 878–890.

www.ingramcontent.com/pod-product-compliance
Lightning Source LLC
Chambersburg PA
CBHW061446180526
45170CB00004B/1579